CAMBRIDGE GEOGRAPHICAL STUDIES

8. MODELS OF SPATIAL PROCESSES

An approach to the study of point, line and area patterns

CAMBRIDGE GEOGRAPHICAL STUDIES

1 *Urban analysis*, B. T. Robson

2 *The urban mosaic*, D. W. G. Timms

3 *Hillslope form and process*, M. A. Carson and M. J. Kirkby

4 *Freight flows and spatial aspects of the British economy*, M. Chisholm and P. O'Sullivan

5 *The agricultural systems of the world: an evolutionary approach*, D. B. Grigg

6 *Elements of spatial structure: a quantitative approach*, A. D. Cliff, P. Haggett, J. K. Ord, K. A. Bassett and R. B. Davies.

7 *Housing and the spatial structure of the city: residential mobility and the housing market in an English city since the Industrial Revolution*, R. M. Pritchard

8 *Models of spatial processes. An approach to the study of point, line and area patterns*, A. Getis and B. N. Boots

9 *Tropical soils and soil survey*, Anthony Young

Models of Spatial Processes

AN APPROACH TO THE STUDY OF POINT, LINE AND
AREA PATTERNS

ARTHUR GETIS
Rutgers University

BARRY BOOTS
Wilfrid Laurier University

CAMBRIDGE UNIVERSITY PRESS

CAMBRIDGE

LONDON - NEW YORK - MELBOURNE

CAMBRIDGE UNIVERSITY PRESS
Cambridge, New York, Melbourne, Madrid, Cape Town, Singapore, São Paulo, Delhi

Cambridge University Press
The Edinburgh Building, Cambridge CB2 8RU, UK

Published in the United States of America by Cambridge University Press, New York

www.cambridge.org
Information on this title: www.cambridge.org/9780521103541

First published 1978
This digitally printed version 2009

A catalogue record for this publication is available from the British Library

Library of Congress Cataloguing in Publication data

Getis, Arthur, 1934–
Models of spatial processes.

(Cambridge geographical studies; 8)
Bibliography: p. 183
Includes index.
1. Geography–Statistical methods. 2. Spatial analysis (Statistics) I. Boots, B. N.,
joint author.
II. Title. III. Series.
G70.3.G47 910'.01'8   75-17118

ISBN 978-0-521-20983-0 hardback
ISBN 978-0-521-10354-1 paperback

To our parents

CONTENTS

Figures ix   Tables xi   Preface xiii

1	An introduction to spatial processes	1
1.1	Spatial processes	1
1.2	A framework for viewing spatial processes	4
1.3	Models of spatial processes	9
1.4	Some illustrations of the utility of pattern analysis	11
1.5	Analytic procedures involved in pattern analysis	14
1.6	The meaning and use of some important terms	15
1.7	The literature in this area	16
2	**Point patterns: Poisson process model**	**18**
2.1	Poisson process model	18
2.2	Pattern measurement	20
2.3	Truncated Poisson process model	35
3	**Point patterns: mixed Poisson process models**	**38**
3.1	Additive processes	38
3.2	Multiplicative processes	48
	Appendices to chapter 3	68
4	**Truly contagious models, disturbed lattices and information theory**	**71**
4.1	Truly contagious models	71
4.2	Disturbed lattice models	78
4.3	Pattern models and information theory (quadrat approach)	81
5	**Line patterns**	**86**
5.1	Processes for generating line patterns	86
5.2	Properties of line patterns	92
5.3	Path models	93
5.4	Tree models	99
5.5	Circuit models	104

5.6	Cell models	107
5.7	Two-phase mosaics	116
	Appendices to chapter 5	118
6	**Area patterns: the cell model**	121
6.1	Area-generating point processes	121
6.2	Properties of area patterns	123
6.3	Processes generating contiguous patterns	126
7	**Area patterns: the Johnson—Mehl model**	145
8	**Area patterns: clumping models**	152
8.1	The Roach approach	152
8.2	The Getis—Jackson model	156
	Appendix A: Introduction to probability theory	165
A.1	Set notation	165
A.2	Combinatorial theory	166
A.3	Binomial coefficients	169
A.4	Multinomial coefficient	170
A.5	Occupancy theory	171
A.6	Probability distributions	171
A.7	Moments of probability distributions	177
	Appendix B: Table of critical values of D in the Kolmogorov—Smirnov one-sample test	180
	Glossary of general notation	181
	Bibliography	183
	Index	195

FIGURES

		page
Fig. 1.1	A diagram for viewing spatial processes	3
1.2	Sector notation added to fig. 1.1	4
1.3	Patterns resulting from various spatial processes	6
1.4	The arrangement of points relative to each other and to the study area	7
1.5	An hexagonal net	8
2.1	Pattern produced by Poisson process (n = 200)	19
2.2	One hundred points and twelve circular quadrats placed by a Poisson process	21
2.3	Fifty points placed by a Poisson process in an area divided into a lattice of twenty-five square quadrats	22
2.4	Only in the inner bounded area is there confidence that the Poisson process model holds	28
2.5	Since B is closer to A than C the measurement BA should be included in the sample, but not AB	28
2.6	No measurement would be taken from point A since it is closer to the boundary of the study area than it is to its nearest neighbor B	28
3.1	Patterns resulting from additive processes	39
3.2	Pattern produced by superimposing two point patterns	40
3.3	Pattern resulting from heterogeneous point process	46
3.4	Patterns resulting from multiplicative processes	49
3.5	The gamma distribution	51
3.6	Towns of population greater than 1000 in Pennsylvania in 1970 by quadrats	54
3.7	Pattern resulting from Neyman Type A process	61
3.8	Pattern resulting from Polya–Aeppli process	69
4.1	Triangular lattice disturbed by 0.2	80
4.2	Distributions representing minimum and maximum values of H	83
4.3	Contribution to H for values of p	83
5.1	Basic assumptions of line-generating processes	87
5.2	Point and line sets	87
5.3	Contrasting P assumptions of two line-generating processes	87
5.4	Processes illustrating simultaneously and sequentially generated lines	88
5.5	Line patterns generated using different circuit forming assumptions	89
5.6	Line pattern generated by restrictions on the number of lines incident at a point	89
5.7	Line patterns resulting from processes involving different combinations of basic assumptions	90

5.8   Line patterns resulting from processes involving different combinations
      of basic assumptions   91
5.9   Line-generating process involving changing assumptions   92
5.10 Two different paths across a rectangle   94
5.11 Probability density function for the lengths of paths across a
      rectangle   95
5.12 Selected paths across a circular plaza   97
5.13 A tree-generating process   100, 101
5.14 An exodic tree   102
5.15 Reduction of a circuit network to tree form   103
5.16 A circuit-generating process (E = 2.89)   104
5.17 A circuit-generating process (E = 5.41)   105
5.18 Circuits produced by interaction of itinerant merchants   106
5.19 A line defined in terms of polar coordinates   107
5.20 A cell-generating process   108
5.21 A pattern produced by a cell-generating process   109
5.22 Definition of nearest neighboring lines   113
5.23 Line patterns resulting from different generating processes   114
5.24 A random walk technique for sampling line patterns   115
5.25 A two-phase mosaic   116
5.26 The nearest neighbor technique for a line pattern   118
5.27 The random walk–reciprocal nearest neighbor technique for
      a line pattern   120
6.1   Basic assumptions of area-generating processes   122
6.2   Processes illustrating simultaneously and sequentially generated
      areas   122
6.3   Contrasting assumptions concerning overlapping   123
6.4   Area patterns resulting from processes involving different combinations
      of basic assumptions   124
6.5   Area-generating process involving changing assumptions   125
6.6   An area pattern generated by the cell model   127
6.7   The incidence of four edges at a vertex in $\mathcal{T}$ 129
6.8   Location of sample cell patterns   133
6.9   Sample cell patterns   134
6.10 Pattern produced by a modification of the cell model   137
6.11 $\mathcal{D}$ for the pattern illustrated in fig. 6.6   139
6.12 Probability density function of a randomly selected Delaunay
      triangle angle   140
6.13 Point pattern analyzed by the use of $\mathcal{L}$ 142
6.14 An area pattern generated by the cell model for use in information
      theory analysis   143
7.1   The Johnson–Mehl model for times (a) t = 1, (b) t = 3, (c) t = 7   145
7.2   An area pattern generated by the Johnson–Mehl model   148
8.1   An area pattern generated by the clumping model   153
8.2   A modified clumping model   162
A.1   Venn diagram containing subset A   166
A.2   Venn diagram showing overlapping subsets   166
A.3   Venn diagram showing an intersection and a union of subsets   166
A.4   Tree diagram   168
A.5   Binomial distribution (n = 7, θ = 0.6)   175
A.6   Steps in the construction of a probability model   177

TABLES

Table		page
2.1	Frequency of points per circular quadrat shown in fig. 2.2	22
2.2	Poisson process model: observed frequency of points taken from fig. 2.3 and expected values	24
2.3	Expected mean distance and standard deviation for nearest neighbors 1–5 in a Poisson process generated spatial pattern	31
2.4	Observed and expected mean distances to first five nearest neighbors for the pattern shown in fig. 2.3 (in kilometers)	31
2.5	Observed and expected proportion of reciprocal pairs, first to fifth nearest neighbors (based on fig. 2.3)	32
2.6	Truncated Poisson process model: group size in free play of four-year-olds at Cambridge, Mass., nursery school	36
3.1	The addition of two Poisson random variables: observed and expected values	41
3.2	Poisson plus Bernoulli process model: observed and expected frequencies	45
3.3	The double Poisson process model: observed and expected frequencies based on pattern shown in fig. 3.3	48
3.4	Compound negative binomial process model: observed and expected frequencies of the location of towns of 1000 population or more in Pennsylvania in 1970 by quadrats	53
3.5	Neyman Type A process model (Poisson–truncated Poisson process): observed and expected frequencies based on simulated data shown in fig. 3.7	60
3.6	Neyman Type A process model: observed and expected frequencies for the location of acceptors of tuberculosis control in cattle (Sweden), 1900–1924	62
3.7	Logarithmic series process model: observed and expected frequencies of clusters of houses in Piscataway Township, New Jersey in 1850	65
3.8	Polya–Aeppli process model: observed and expected frequencies based on pattern shown in fig. 3.8	69
4.1	Mean distance and variance of nearest neighbors in disturbed triangular lattice	80
4.2	Information measure (H) for pattern shown in fig. 2.3, for Poisson process model, and for a random unspecified probability function approximately equal to a maximum (based on 50 points and 25 quadrats)	84

5.1	Frequency distribution of path lengths	98
5.2	Size frequency distribution of marketing systems	107
5.3	Frequency of intersection of an arbitrary line of unit length with $\mathscr{L}_M$	110
5.4	Moments for properties of polygons in a pattern generated by the Miles model and a comparison with some values obtained from fig. 5.21	111
5.5	Interpoint distances on a traverse	120
6.1	Moments for properties of the cell model	127
6.2	Moments for properties of the pattern illustrated in fig. 6.6	128
6.3	Estimates of higher order moments for properties of the cell model	130
6.4	Additional moments for properties of the pattern illustrated in fig. 6.6	130
6.5	Crain estimates of the frequency distribution of k-sided polygons for the cell model	130
6.6	Crain estimates and the observed frequency of contact numbers for the pattern illustrated in fig. 6.6	131
6.7	Kiang estimates of cell area and observed areas for the pattern illustrated in fig. 6.6	131
6.8	A test of the cell model using data on bus service hinterlands taken from fig. 6.9	134
6.9	Expected proportions of contact numbers using original and modified Smalley model: tests on basalt cracks at four sites	138
6.10	Density dependent values of $\mathscr{Q}$	141
6.11	Frequency distribution of α for the sample point pattern of fig. 6.13	142
7.1	Moments for properties of the Johnson—Mehl model	148
7.2	Moments for properties of the patterns illustrated in fig. 7.2	149
7.3	Estimates of coefficients of variation and correlation coefficients for properties of the Johnson—Mehl model	149
8.1	Frequency distribution of clump sizes for the pattern illustrated in fig. 8.1	155
8.2	Proportion of study area covered by k circles (based on fig. 8.1)	156
8.3	Frequency distribution of clump sizes for pattern of cities	158
8.4	Frequency distribution of clump sizes for pattern of urban spheres of influence	158
8.5	Observed and expected proportions of study area covered by k pollution zones	161
A.1	Outcomes for rolls of a pair of dice	172
A.2	Binomial distribution for $n = 7$ and $\theta = 0.6$	174
A.3	Examples of Poisson distributions, $\lambda = 2.4$ and $\lambda = 1.2$	176
A.4	Calculation of μ_1	177

PREFACE

Geographers have always been interested in map patterns. Some have been concerned with the development of techniques to describe and summarize map patterns, thus facilitating their comparison. Others have been interested in map patterns for the evidence they provide in support of hypotheses concerning the way objects become located in space. But geographers are not the only ones interested in such patterns; they are of considerable importance to such disciplines as ecology, biology, geology, forestry, astronomy, and statistics. As a result of this widespread concern the literature available to the geographer is extremely diverse. Our primary goal in this book is to unify much of this literature and present it as a cohesive statement organized around a common theme.

The theme is one which stresses a process-oriented approach, where the processes examined in any context are those suggested by the subject matter of the pattern. The processes examined are stochastic (probabilistic) ones and an inferential approach is pursued. The main goal of the book, however, cannot be achieved without a price. In developing our own perspective we present some work in a form other than that conceived by the original author. In addition we have not considered some relevant material. The omission of certain topics does not imply that we consider them unimportant; indeed, several, such as pattern recognition and spectral analysis, are very interesting areas of study with considerable potential but they are not suited to the probabilistic approach taken here.

Because we are interested in communicating pattern analysis work to a wide audience we have tried to simplify the material as much as possible. We believe that the shroud of mathematical mystery which surrounds pattern analysis in the minds of many, particularly those not trained in mathematics, deters potential users. Consequently we have attempted to reduce the mathematical component to a minimum by such devices as simplifying expressions (although this may mean they are not presented in their most elegant mathematical forms) and curtailing the inclusion of proofs. The latter appear only when they assist in clarifying the role of the particular model in which they are incorporated. The reader requires only a basic knowledge of statistics and probability. To aid him in this and to provide a brief refresher for those already familiar with such material we have included a short appendix which outlines the basic elements needed to understand the material in the book. If desired, the appendix can be supplemented by reference to the standard texts by Feller (1950), *An Introduction to Probability Theory and Its Applications,* Vol. 1, 1950 and Freund (1971), *Mathematical Statistics.* To further facilitate comprehension we have included a large number of diagrams and tables, many referring to actual examples of pattern analysis. Further in this regard, the bibliography contains reference to pattern analysis for various phenomena in several disciplines.

We should stress that while it is our wish to unify as much work as possible within our framework the book is not intended to be an exhaustive statement. Instead, it represents our own particular approach to pattern analysis and we recognize that alternatives exist and that our approach may not always be the one preferred by any individual reader. In any book decisions have to be made on what material is essential to the goal of the book and must be included, and which material is at best peripheral and may be excluded without detriment. Such decisions can be made arbitrarily, but we made ours with two major criteria in mind. First, we attempted to determine whether the work fell within both the intended philosophical and methodological scope of the book. There is a temptation to make a book such as this all inclusive, especially when there are few other works in the same field. Our main concern was to create a study which was as catholic as possible and so we have stressed the inclusion of the universal and often excluded the specific. Second, our choice of material was guided by the availability of other sources. Where texts already exist we have tried to avoid replicating the material. For instance, more material on traditional network studies could have been included, especially in the chapter on line patterns, but much of this material is already available (e.g. *Network Analysis in Geography* by P. Haggett and R. J. Chorley, 1969).

One of the effects of our concern with general models has been the limited discussion of simulation approaches. Many simulation studies are concerned with attempts to model specific instances of reality, often because the empirical circumstances are too complex to be handled by analytic methods. Instead, we prefer to present general models and to illustrate how these might be modified to deal with particular empirical instances. Also, since we have chosen to emphasize a particular approach to pattern analysis, viz. by way of concentrating attention on the processes responsible for the patterns, alternative approaches receive only brief consideration.

Another result of the general approach taken is that the models we present are predominantly abstractions. That is, in terms of the models themselves, the elements involved are just points, lines or areas. We have followed this format so that the reader may accomplish the necessary interpretation, identifying points as cities, churches, nest sites, or whatever, lines as roadways, footpaths, tracks, etc., and areas as market areas, ecclesiastical parishes, animal territories, and so on, depending on his particular empirical circumstances. We present only a cursory treatment of problems which arise after the elements in the model have been identified. The most severe problems involved here are those of spatial autocorrelation, but a useful source is available in *Spatial Autocorrelation* by A. D. Cliff and J. K. Ord (1973).

Those named in the bibliography are the writers who provided the intellectual foundation for this work, and to them go our deeply felt thanks. It will be noted that M. F. Dacey is referred to very often there and in the text. It is to him that we owe so much for our interest and understanding of this area of research. In addition, J. K. Ord, J. A. Whitney, and J. I. Naus, through their comments and criticisms contributed considerably to the quality of our work. P. Haggett, our colleague and teacher, stimulated us greatly through his extensive work in spatial patterning. B. H. Farmer and E. A. Wrigley did much to encourage and guide us in the preparation of the book for press. For the help they have given us we are indebted to the staff of Cambridge University Press and Ronald Foresta of Rutgers University. We would also like to thank the Rutgers University Research Council for their generous financial support for this project.

Finally, we would like to take this opportunity to thank our respective wives, Judy and Chris, for the help and support they gave us in various tasks throughout the preparation of this book.

Arthur Getis
Barry Boots

June 1977

Acknowledgements

Thanks are due for permission to reproduce some figures and tables from other publications:

Figs 1.1, 1.2 and 1.3. Copies of figures 1, 2 and 3 respectively, p. 77, 78 and 79 from Arthur Getis, 'Representation of Spatial Point Pattern Processes by Polya Models,' in M. Yeates (ed.) *Proceedings of the 1972 Meeting of the IGU Commission on Quantitative Geography*, McGill–Queen's University Press, Montreal and London (1974).

Figs 5.10 and 5.11. Copies of figures 1 and 4, p. 170 and 174 from Maurice Horowitz, 'Probability of Random Paths Across Elementary Geometrical Shapes,' *Journal of Applied Probability*, 2, 169–177 (1965).

Fig. 5.15. Copy of figure 15.28, p. 660 from Richard J. Chorley and Peter Haggett (eds.) *Models in Geography*, Methuen and Co. Ltd, London (1967).

Fig. 6.12. Copy of figure 3, p. 112 from R. E. Miles, 'On the Homogeneous Planar Poisson Point Process,' *Mathematical Biosciences*, 6 (1970), 85–127. (American Elsevier Publishing Co., Inc.)

Fig. A.6. Copy of figure 3.2, p. 88 from Donald L. Harnett *Introduction to Statistical Methods* (Second Edition), Addison-Wesley Publishing Company, Reading, Mass. (1975).

Appendix B. Table of critical Values of D in the Kolmogorov–Smirnov One-sample Test. Adapted from Massey, F. J., Jr (1951). 'The Kolmogorov–Smirnov test for goodness of fit,' *Journal of the American Statistical Association*, 46, 70.

1
An introduction to spatial processes

Basic to the field of geography is the map study of the spatial patterns of many kinds of objects. In fact, the subject matter explored by geographers is often chosen only if it is important to study its spatial arrangement. It is no accident that geographers study settlements, land uses, river systems, and flows of goods, ideas and people. Much can be learned about these subjects by testing hypotheses about their patterns, usually as represented on maps. Very often the hypotheses are deduced from theories about physical, economic, social and political forces. These forces are instrumental in bringing about changes. It is the spatial aspects of these changes that we call spatial processes. There are various spatial processes but for the moment, to fix ideas, it is helpful to think of them as tendencies for elements to come together in space (agglomeration) or to spread in space (diffusion).

The approach we employ in this book involves identifying a number of spatial processes, suggesting appropriate mathematical models for them, and examining the spatial arrangements that arise from the processes.

The ultimate goal is to explain observed map patterns. The models, which represent spatial processes, yield various arrangements of objects on maps. These theoretical patterns are then compared to observed patterns and conclusions are made about the spatial processes. This three-step procedure—process, model, pattern—is repeated for many situations. To introduce our approach to map pattern analysis, we begin with a discussion of what we mean by 'spatial process'.

1.1. Spatial processes
From popular usage it is clear that the term 'process' implies a sequence of events, which may be either discrete or continuous. But it is not just any sequence of events. There is the implication that the sequence is carried on in some definite manner which leads to some recognizable result. Also implied is the notion of change, the transformation of a set of parts into a finished whole, the creation of a new form in an environment in which it had not existed previously. These changes are the result of certain forces,

be they physical, social, economic or psychological. In the analysis we pursue here we are not directly interested in the forces *per se*. Rather, our attention is focussed on spatial aspects of these processes, and the patterns to which they give rise.

Since an adequate explanation of process involves using such terms as 'sequence' and 'change', any understanding of spatial process necessitates the inclusion of a time element either explicitly or implicitly, although a consideration of time may be incidental or unimportant. This is because time is best conceived of as a relative rather than an absolute quantity.

Spatially we might envisage an observed pattern as being a single frame from a motion picture reel. Our goal would be to identify the responsible spatial processes so that given any frame of a film we could predict adjacent frames. Such a level of sophistication has rarely been achieved even in the most simple circumstances, but we shall proceed as though this goal is ultimately attainable. In some films each frame is different from the preceding one and in order to comprehend the story it is necessary to view the complete film. For example, this would be the case if we represented each player in a football game as a point and watched the movement of those points over the course of a game and then attempted to make sense of the changes in pattern that we observed. In contrast, there are other films in which there are long sequences of identical frames and a set of judiciously selected stills might be sufficient to enable us to understand what had taken place. Thus, a film illustrating boundary changes in Western Europe in the present century might well be summarized by stills taken in 1900, 1914, 1919, 1939 and 1945. Much will depend on the speed at which the film is run. For instance, we may speed up the film so that the changes occur rapidly relative to the playing time of the film. This is what is meant above by the relative nature of time.

If we continue our cinematic analogue a little further, we might identify other kinds of films. There are those in which the major portion of the action occurs in the first few frames and for which the overall plot can soon be discerned. For example, once electoral districts are established they may undergo only a few minor adjustments. To note a further example, a force (e.g. unsatisfied demand) may cause the siting of new central places in an area until the density of the central places reaches a critical level, at which time no more unsatisfied demand remains and the pattern stabilizes (until new changes occur in the level of demand).

In other films some sequences are repeated, often according to regular cycles. A film of the journey-to-work behavior of individuals (represented as points) over a week's time might contain such repetitions.

Finally, the most complex films will be those in which there are no repeated frames and for which the story goes on in 'soap opera' fashion. One example would be a film in which we represented each person in a city as a point and then observed the movement of the points over time. Not only are individuals (points) in a continual state of flux, but new points are also constantly entering and leaving the environment by way of migration and birth–death mechanisms.

Many of the difficulties encountered in pattern analysis stem from the fact that, at best, we have but a few of the frames from any film, and more often we have only an isolated frame. From our analogue it will be apparent that whether these frames can be used successfully in pattern analysis will depend a great deal on the overall nature of the film, that is, on both the real-world subject matter and the spatial process operative on it.

Consequently, we shall be primarily concerned with those instances in which the spatial processes involved are relatively simple, not only in their spatial manifestation, but also in the extent to which the time element is explicitly considered. In many of the processes we shall encounter, the time element is unimportant in that the patterns we examine are assumed to develop 'instantaneously' or over some unspecified time horizon.

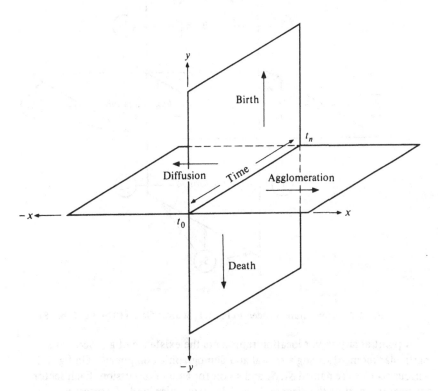

Fig. 1.1 A diagram for viewing spatial processes. Source: Getis (1974, fig. 1, p. 77)

1.2.   A framework for viewing spatial processes

Fig. 1.1 will help to clarify our view of spatial processes. It is an impressionistic representation of the spatial and demographic processes responsible for pattern change. Economic, political or social system forces are not considered. Only points are used in our illustrations, and not other geometric forms such as lines or areas. t_0 represents the earliest time period and t_n the final time period. The horizontal axis (x) represents the diffusion–agglomeration continuum. Positive x values indicate an agglomerative spatial process. Agglomerative tendencies become stronger as x increases. A diffusion force may exist where x is positive but the agglomeration force is greater. In the same manner a positive y value implies that more births than deaths are evident. A pure birth process would take place at the highest y value.

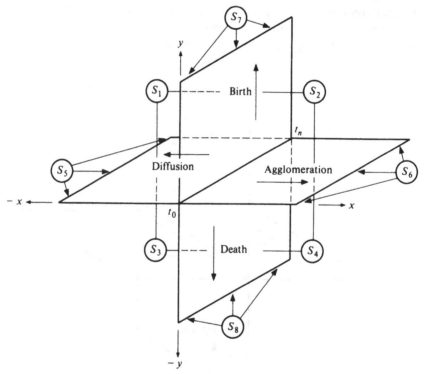

Fig. 1.2 Sector notation added to fig. 1.1. Source: Getis (1974, fig. 2, p. 78)

A point at any x, y, t location represents the existence of a process at a particular moment having a spatial and demographic component. On fig. 1.2 various sectors are named S_1, S_2 and so on for ease of discussion. Each sector represents a part of the diagram in which a particular kind of process is evident. Note that by looking at a given sector we are assuming that x and y

are held constant over time. If x and y were allowed to vary in different ways over time an infinite variety of processes could be depicted. Here just eight kinds of processes are shown. Some of the possible point patterns resulting from the eight processes are displayed in fig. 1.3.

Sector S_1 $(-x, y)$ represents a process where births are more in evidence than deaths and diffusion tendencies are greater than agglomerative forces. There are at least two kinds of patterns which might result from such a situation. The first, S_{1a}, is the 'contagion' effect, where at each succeeding time period more objects are noted in locations farther and farther from the original objects. Examples are epidemics, settlement development, or the diffusion of information. In S_{1b} there is a relocation of objects during each time period. Migration is the most characteristic process. Because there is an increase in objects at each time period the process may aptly be called the 'pied piper' effect—a growth and relocation situation.

S_2 (x, y) represents a 'family development' effect. Where birth and agglomeration forces are greater than death and diffusive forces, there is an increase in numbers among the already existing objects. Examples of this would be the attraction of certain objects for other objects such as the agglomeration of industry or of mutually supportive species.

The 'survival of the fittest' effect of S_{3a} $(-x, -y)$ results from the decrease in objects as diffusion continues. As time elapses objects disappear and a less dense and more diffused pattern results. There are a number of plant species subject to such a process. Among animals the newly hatched turtle is an example. In S_{3b} there is the counterpart of the pied piper effect. This might be called the 'lost cause' effect. An example of this pattern would be unsuccessful migrations.

S_4 $(x, -y)$ represents the decrease in the number of objects as clustering continues. Examples would come from processes where objects are lured to their demise, such as rats tempted into rat traps.

The processes represented by S_5 $(-x, 0)$ and S_6 $(x, 0)$ exhibit no increase or decrease of objects over time. The spread and migration processes and the 'town meeting' effect take place without change in numbers. Situations such as these, where movement is considered without population change (the entire x, t plane), represent pure spatial processes.

In like manner pure population processes would be exhibited as the vertical plane (y, t). At the extremes of the plane S_7 $(0, y)$ and S_8 $(0, -y)$ are pure birth and death processes respectively. In these cases births or deaths may be conceived of as being located by a random process.

An example of a combined agglomeration and diffusion process is the 'friction of distance' effect. Such a situation occurs when there is a strong force for spread and also a strong force for clustering. An equilibrium is reached when the forces balance each other. On fig. 1.1 this situation is found at the intersection of the x, t and y, t planes.

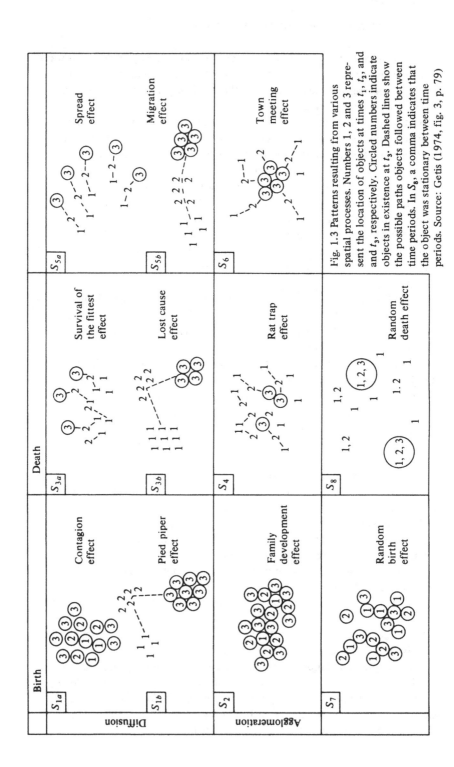

Fig. 1.3 Patterns resulting from various spatial processes. Numbers 1, 2 and 3 represent the location of objects at times t_1, t_2, and t_3, respectively. Circled numbers indicate objects in existence at t_3. Dashed lines show the possible paths objects followed between time periods. In S_8, a comma indicates that the object was stationary between time periods. Source: Getis (1974, fig. 3, p. 79)

On fig. 1.3, observe the pattern of 1's in each diagram. If this were the only information available to us, we can see how confusing it would be to try to predict the patterns for successive time periods. The fact that many of the patterns for t_1, t_2 and t_3 look the same is further evidence of the futility of meaningful analysis without knowledge of patterns at other time periods.

The number of possible patterns would increase if an environmental dimension were added. Environment here is taken to mean the setting within which an object is located. It may have supportive or non-supportive qualities. If

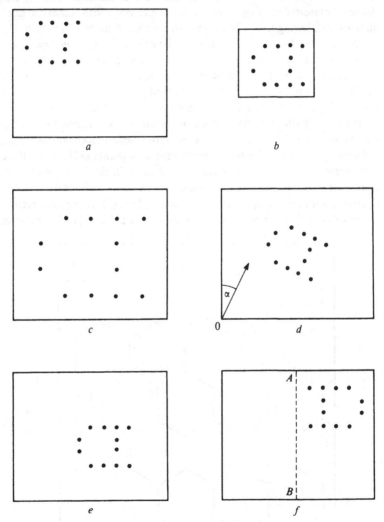

Fig. 1.4 The arrangement of points relative to each other and to the study area

the support is not everywhere equal then the pattern of objects will reflect this heterogeneity. Objects may be barred from some areas and flourish in others.

In map pattern analysis it is the arrangements of geometric objects which we study; that is, the points, lines, and areas which we use to represent the real-world objects. The choice of symbols is the result of the theoretical knowledge which we bring to bear on the problem being studied. These symbols are recognized as possessing certain properties, the most important of which are associated with the dimensionality of the object. Points are considered dimensionless; they have neither length nor width. Lines are one-dimensional, having length but no width, and areas are two-dimensional, having both length and width. Thus, in the case of a city shown on a map as a point, we are representing a complex three-dimensional object by a geometric figure of zero dimensions. Since points are dimensionless an infinite number of them can theoretically be located in a plane, although in the reality of two-dimensional map-making points do consume area.

The word 'pattern' in its purest sense refers to the location of occurrences relative to one another. In this sense, pattern is independent of both scale and density. Consider the simple arrangement of points in fig. 1.4*a*. If we reduce the size of the rectangle (or study area) as in fig. 1.4*b*, we change the density of points per unit area of the study area, but we have done nothing to alter the spatial arrangement of the points. In fig. 1.4*c* we have returned to the original size of study area and have retained the original number of

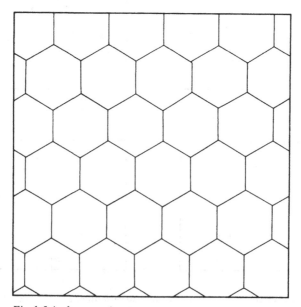

Fig. 1.5 An hexagonal net

points, so that the densities in figs 1.4*a* and 1.4*c* are the same. However, in fig. 1.4*c* we have increased the distance between all pairs of points, but the pattern remains unchanged. In fact, we can subject the points in fig. 1.4*a* to a series of 'transformations' without destroying the pattern. In fig. 1.4*d* we have rotated the pattern through angle α about an arbitrary point 0. In fig. 1.4*e* we have moved each point the same distance in a constant direction. Finally, in fig. 1.4*f* we have repeated the pattern by reflection about the line *AB*. Such a rigid definition of pattern is useful in those instances in which the components of a pattern are seen only as geometric objects, and not as the real-world objects they represent. There is the difficulty that armed only with a pattern of geometric objects, we are unable to make any statements about the processes responsible for that pattern. Consider fig. 1.5. It is a representation of the jointing in columnar basalt, the cells in a honeycomb, and the market areas of settlements in a perfect central-place network. In each instance the processes responsible for the pattern are different and we cannot discover the processes by examining only the single pattern. We must have more information.

1.3. Models of spatial processes

In this book we present a variety of models, which incorporate an equal variety of spatial processes, from the very simple to the quite complex. We feel, therefore, that it would be beneficial to present a general framework for classifying models of spatial processes.

Processes involve the inclusion of time elements, either explicitly or implicitly, and while time is continuous it is often possible to recognize distinct points or stages. Consequently, we can characterize spatial processes in terms of the number of steps involved in them and the nature of those steps.

Since we shall only concern ourselves with point-generated patterns, the patterns of points, lines or areas we study will have been created according to various rules concerning the location of points. These rules represent the assumptions inherent in the spatial process and thus become our spatial-process models. Let us designate the assumptions responsible for the generation of the point elements of the pattern, **P** assumptions. If we are dealing with line or area patterns, at least one more group of assumptions is required to locate the line and area elements of the pattern. Designate these as **L** or **A** assumptions. Of necessity, then, process models generating line and area patterns are made up of a minimum of two groups of assumptions. The line or area-generating process is labeled a **PL** or **PA** process respectively.

Let us consider the nature of **P** and **L** or **A** assumptions more closely, especially in terms of the treatment of the time element. There are some processes in which the time element is immaterial or ignored, giving rise to *instantaneously* created patterns. For **PL** or **PA** processes this kind of pattern would result from the application of:

(1)   **P** assumptions to locate *all* the points at one moment in time and
(2)   **L** or **A** assumptions to generate *all* lines or areas simultaneously with the points.

In contrast to such processes we may recognize those which give rise to patterns which grow or evolve *sequentially* over time. A typical process of this type might result from:

(1)   **P** assumptions to locate a few points;
(2)   **L** or **A** assumptions to generate lines or areas with respect to the points located by **P** in step 1;
(3)   the same **P** assumptions to locate some new points;
(4)   the same **L** or **A** assumptions to generate additional lines or areas;
(5)   repetition of steps 3 and 4 until the final pattern is produced.

If the **P** and, say, **L** assumptions remain the same throughout the process, we can describe the process as

PLPLPL . . . PL

where each **PL** is one sequence of the basic process components. In general, the number (N) of **PL** or **PA** components can be thought to signify the importance of growth elements (and thus time) to the process. In this way the simple **PL** or **PA** process involving the instantaneous creation of pattern might be considered a special case in which $N = 1$.

It is interesting to consider briefly how N is determined in sequential patterns. One way is *external* to the evolving pattern; the process may cease after some given number of specified time intervals. Alternatively, N may be determined *internally* by the nature of the pattern which results. For example, the process might terminate when the number of newly created regions (areas) reached a specified level, or when a certain portion of the plane had been colonized. In such cases it would be difficult to specify N in advance.

It should be apparent now that the designation of processes in terms of assumptions enables us to categorize more complex patterns. In addition, for line and area patterns we have a means of creating new models from simple basic processes. In complex sequential processes both the **P** and **L** or **A** assumptions change over time. Indeed, in many geographic systems the process is being constantly modified over time, with early-established elements of the pattern often exerting considerable influence on subsequently-established elements. Consequently, a procedure in which the process for locating the points remains constant through time but in which a new method of generating lines or areas occurs through time would have the form

$$P_1L_1P_1L_2P_1L_3 \ldots P_1L_n \text{ or } P_1A_1P_1A_2P_1A_3 \ldots P_1A_n$$

In contrast,

$$P_1L_1P_2L_1P_3L_1 \ldots P_nL_1 \text{ or } P_1A_1P_2A_1P_3A_1 \ldots P_nA_1$$

would be a process in which the points are located according to changing assumptions while the method of generating lines or areas remains constant.
Finally,

$$P_1L_1P_2L_2P_3L_3 \ldots P_nL_n \text{ or } P_1A_1P_2A_2P_3A_3 \ldots P_nA_n$$

would signify a process in which both the means of generating the points and the lines or areas change through time. These may be processes in which the assumptions are independent of each other, or processes in which a set of assumptions is dependent on one, some, or all of the previous assumptions.

Since we are dealing in this book with point-generated spatial patterns, every one of our processes involves at least one **P** group of assumptions. The simplest meaningful process we encounter consists of just two **P** assumptions, resulting in an instantaneously created point pattern. In chapters 2, 3 and 4 various models with different **P** assumptions are discussed. In chapter 5 concern is for models of the form **PL** and in chapters 6, 7 and 8 **PA** models are explored.

In the preceding paragraphs we have developed a scheme which may imply that in what follows a large variety of models will be discussed. This is not quite the case. Current research allows us to write about only the simplest models. Very little research on patterns considers complicated assumptions dealing with points, lines and areas. The framework developed in this introduction is just that, a way of viewing pattern-producing models. Perhaps this framework will stimulate the development of more elaborate models.

1.4. Some illustrations of the utility of pattern analysis

Some general examples of the possible applications of point-generated pattern analysis follow. They serve to illustrate the flexibility of pattern analysis. This is a suggestive rather than an exhaustive list.

1.4.1. Patterns of points (P models)

Agglomeration or grouping. Suppose theory suggests that a particular set of objects (plants, animals, people, towns, etc.) tends to group or agglomerate in certain ways. Point pattern analysis is helpful in measuring various characteristics of the groups (size, spacing, density, etc.) and leads to the testing of hypotheses derived from theory. For example, studies of animal behavior suggest that certain types of spatial patterns help to verify theories of territoriality and social organization.

Diffusion. Many theories have been proposed for the way individuals or ideas spread or spatially multiply. Point pattern analysis can be helpful in verifying the existence of a diffusion process and in calibrating rates of change. An example comes from the study of the spread of information. Various theories imply that new ideas spread according to principles based on the nearness of possible communicants and their resistance to accepting ideas. By the analysis of patterns at various moments in time and in different environments, these notions can be tested.

Competition. It is often desirable to investigate spacing characteristics when it is suspected that competitive forces are at work. Sometimes competition yields maximum spacing and other times grouping. A well-known example comes from the literature on town spacing. Spatial aspects of economic theories of marketing can be tested by point pattern analysis. There are also theories of plant and animal competition.

Segregation or association. Hypotheses about the existence of spatial segregation in a many-species population of individuals can be tested with point pattern analysis. Following urban rent theory we may expect two kinds of land uses to 'repel' each other. This expectation can be tested, as well as theoretical expectations of an association among several land uses.

Pattern change. Many theoretical statements deal directly with the manner in which patterns change. For example, the birth and death processes of plant and animal populations as well as human populations may very well be studied by point pattern analysis. Interest might be in rates of change in patterns.

1.4.2.   Patterns of lines (PL models)

Point-generated line patterns are the result of various kinds of *linkage* processes. Usually attention focusses on the number of points that are linked and the manner in which the linkage is achieved.

Paths. The simplest form of linkage takes place when just two points are joined. We can study aspects of the form and length of the paths which result. We might have data on point-to-point vehicular movements in cities, maze-running behavior of laboratory rats, or pedestrian paths across city plazas and parks.

Branching. With branching patterns more than two points are linked but no closed loops are created; that is, there is only one path between any two points. Branching networks occur throughout nature, as for example in rivers and trees. Sociological power structures, decision-making hierarchies, and some transportation networks are examples of branching in man-made structures. Properties of such patterns can be compared to those which result from process models.

Circuits. If we relax the constraint applied in the creation of branching patterns by permitting the existence of closed loops, we can generate circuits. Most transportation networks are of this form. By examining characteristics of observed circuits we can again test hypotheses concerning the processes that lead to such patterns.

Cells. Finally, if we impose the constraint that the linkage process must result exclusively in closed loops, we have created a cellular network. Field patterns, politico-administrative networks of many kinds, and some fracture patterns in rock are relevant examples. Two patterns result from the creation of a cellular network. One is the line pattern formed by the network of cell boundaries (edges) and the other, the complement of the first, is composed of the areas defined by the line pattern. Depending on the particular empirical circumstances, characteristics of either or both patterns can be explored.

1.4.3. Patterns of areas (PA models)

Coverage. There are many instances in which a point acts as a node from which a surrounding area is serviced. Central places provide goods and services for adjacent rural areas, clinics provide health care for urban neighborhoods, radio and television stations broadcast over a portion of a state, water sources provide drinking water for dispersed populations, and so on. Often more than one node provides services for a particular area, in which case there are overlapping service areas. Characteristics of the resulting area patterns (such as size and shape) can be compared with those of patterns which result from the operation of various models incorporating area-generating processes.

Assignment. In what may be regarded as the inverse process of coverage, areas may be assigned to a particular point. In many politico-administrative systems voters are assigned to a particular polling place, passports must be obtained from a single office, taxes paid to a specific revenue office, and so on. If the assignment process is a deliberate one, mutually exclusive areas are usually created, and the overlap of areas which can occur with the coverage process is absent. Once again, models incorporating various assignment assumptions can be used to determine the nature of the processes responsible for areas of the kind seen in observed patterns.

Growth. Area patterns can result from the spread of influence away from a set of original point sources. In a region of pioneer development new settlements can be introduced, and, once successfully established, these settlements may become central places which in turn influence the surrounding areas, establishing tributary trade areas. These trade areas might well grow through time as the importance of the central place grows. Some nation states were

created by the gradual expansion of a power base away from a single focal
point. Similarly, some gregarious animal species may extend their territories
if the size of the group increases.

Partitioning. A single undifferentiated plain or study area can be subdivided
by the introduction of a set of internal boundaries. A city may be divided into
census tracts, an uncolonized third of a nation into a grid of township and
range, a conquered nation into four zones of occupation, a basalt flow into
columnar blocks, and so on.

1.5.   Analytic procedures involved in pattern analysis

The purpose of this section is to alert those new to this area of
inquiry that one cannot expect to use pattern analysis with any success
whatever unless one strictly follows the scientific method. In many areas of
inquiry the scientific method is followed loosely, sometimes for great intellectual gain, but in this area of inquiry no such freedom is available. One
must carefully develop his ideas about patterns before he begins to measure
them. It is virtually impossible to develop or test theory by starting with
the pattern itself.

This circumstance is a function of the limitations of the things that are
studied – points, lines and areas. While it is possible to differentiate patterns
by measurements of various kinds, no single measure can describe a pattern
to the degree where slight differences imply different underlying causal
factors. Even when a pattern conforms to expectations one must be careful
to refrain from concluding that the presumed theory or hypothesis holds,
because some other causative variables might have been responsible for
the pattern. Only by testing a single hypothesis in a number of instances or
contexts can one begin to verify presumed notions.

What follows is a brief review of the steps in testing an hypothesis
dealing with map patterns.

(1) *Theory-hypothesis.* Within the framework of the subject matter, select
or develop a theory which needs testing. Since our concern is with patterns
of points, lines and areas, the theory should either refer explicitly to
patterns or be susceptible to modification so that it has a spatial connota-
tion.

At this point, hypotheses about pattern or pattern change can be developed.
These statements consider the way one would expect the properties of the
pattern to behave given the theory. The hypotheses may deal with any of a
large number of pattern characteristics, such as spacing and density.

(2) *Process-model.* List the assumptions which you believe constitute the
theory, paying special attention to those that have spatial meaning. The group

of assumptions presumed to be responsible for the expected pattern is the model. For example, suppose a theory calls for individuals to organize into groups whose spacing depends on the number of individuals in each group. The hypothesis might be derived from the expected mean distance between groups of certain sizes and/or the size of the territory surrounding groups of specified sizes. Suppose it is thought that the procedure by which the groups and their spacing evolved followed a sequence of steps where each new group location was a function of the location of other groups. We would use a P-type model which as closely as possible replicated that presumed process. A special model might be developed for this test or an already existing model used. As we shall see, there are a series of well-known probability models which satisfy the assumptions needed for modeling many map patterns.

(3) *Measurement-testing.* Having developed or chosen an appropriate model we can now determine the kinds of map patterns it will yield. We explore the model's mathematical characteristics to see what is the likelihood of the various possible pattern outcomes. Outcomes usually take the form of a set of expected frequencies which are then compared to a set of frequencies obtained from the actual pattern.

Measuring the pattern in order to obtain the frequency of observed values may be simple or fraught with difficulties. In any case, the measurements derived from the model and the observed pattern are subjected to a statistical test. The chi-square is a popular non-parametric test for comparing a set of observed frequencies with those expected from a hypothesized model.

(4) *Conclusions.* The fact that the theoretical and observed patterns agree does not necessarily prove the theory correct. Reservations about our conclusions will include our degree of confidence in the assumptions selected (model), our concern that other valid assumptions might yield a similar pattern, and our knowledge of the errors that can arise from the statistical procedures used.

1.6.   The meaning and use of some important terms

1.6.1.   Random

Probably as important as any concept in the study of map patterns is that of randomness. The literature is filled with such terms as 'random sampling', 'random variable', 'random pattern', and 'random distribution'. The word 'random' can be misleading. We shall use it only to describe certain sampling procedures (random sampling and random numbers) and certain variables (random variables).

Although 'random' has often been used to describe map patterns that are neither clustered nor even, we refrain from such usage. It is misleading because it implies that a random process has given rise to a pattern that appears haphazard, whereas this may not have been the case at all. For example, suppose that a pattern of town locations is neither clustered nor even. By designating the pattern random we might imply that the process responsible for the creation of the towns was random. Most people would agree, however, that the process of town location is not random.

1.6.2.   Distribution

The word 'distribution' is used often in this book. Since it has several meanings and there is the possibility that one context may be mistaken for another, our convention is to use it technically for one purpose only: to describe a set of values of a variable by associating a relative frequency with each value. As an example, the normal curve represents a particular distribution, one which has the greatest relative frequencies associated with the central values. Other distributions that are spoken of are the binomial distribution, Poisson distribution, and the negative binomial distribution.

The context which we will carefully avoid is that pertaining to the location of objects. When we speak of symbols on maps located in a particular way we use the term 'pattern' rather than 'distribution'.

1.7.   The literature in this area

The work done on pattern analysis is concerned with specific theoretical or empirical problems. Typically, theoretical work deals with the nature of processes and the way patterns are interpreted. M. F. Dacey has been very active in this field and reference to his work will be found throughout this book. The empirical work has to do with the construction of various pattern measurements and the problems that are entailed in their use. There are a few general works that treat pattern analysis as a unified whole. Pielou (1969) devotes attention to both the study of patterns and the ecological problems of species abundance. Rogers (1969*a, b,* 1971, 1972), in a series of articles and now in a book (1974), discusses most of the mathematical derivations and sampling problems of point-pattern analysis, while Haggett and Chorley (1969) cover most of the work on patterns of lines. Matérn's (1960) book on spatial variation is a sophisticated approach to many spatial pattern problems.

In geography much of the pattern literature is briefly reviewed in King (1969), and both Curry (1964) and Hudson (1969) theorize about the study of patterns as the study of the location of settlements. Olsson (1965) and Brown (1968) summarize a good deal of this literature. Others active in these areas will be mentioned in particular contexts later.

In statistics the fundamental work in the probability theory needed for pattern analysis can be found in Ord (1972), Feller (1968), and Johnson and Kotz (1969). There are several collections of papers which give a very good idea of the kinds of spatial subjects treated by a wide variety of statisticians and statistically-oriented disciplinary practitioners. One three-volume set has been edited by Patil (1970) and another by Patil, Pielou and Waters (1971). Papers found in these collections will be mentioned from time to time.

Sometimes it is necessary to refer to tables: a particularly helpful book is that of Abramowitz and Stegun (1965). Very often we refer to the chi-square and Kolmogorov–Smirnov goodness-of-fit tests. We presume that most readers have easy access to a chi-square table; we have provided a Kolmogorov–Smirnov table in Appendix B.

2

Point patterns: Poisson process model

The models that explain patterns of points on maps are called point process models. Using the nomenclature developed in chapter 1, these are classified as models having **P** assumptions. In chapter 2 our task is two-fold: first, to explain the assumptions which underlie the Poisson point process model and, second, to provide ways in which one might identify the patterns generated by the model. The assumptions concern the rules developed for placing points on maps. These should correspond to the theory implicit in the subject matter under study. The identification procedures are varied and sometimes complex. A large literature has been developed on the characteristics of point patterns; only the essential aspects will be discussed here.

For the material of chapter 2 and subsequent chapters to be meaningful it is necessary for the reader to have a grasp of basic probability theory. Appendix A contains a discussion of the rudiments of probability theory, emphasizing spatial applications.

2.1. Poisson process model

The Poisson process is basic for the development of most point pattern models. The assumptions which represent a Poisson process are:

(1) n points are placed in a region where each possible location for a point is *equally likely* to be chosen; and

(2)   the location of each point is *independent* of the location of any other point, i.e. in no way does the selection of a location for one of the points bear on the selection of a location for any other point.

These assumptions can be envisaged by assigning a code number to each location. The code numbers are sampled so that each has an equal chance of being selected. One is chosen and a point placed at the indicated location. This procedure is repeated n times. If a particular code number is redrawn, another point is placed at the indicated location. In the next chapter we will add further assumptions in order to develop more complicated models.

Mathematically, the Poisson distribution is given as

$$P(x;\lambda) = \frac{e^{-\lambda}\lambda^x}{x!} \quad \text{for } x = 0, 1, 2, \ldots \tag{2.1}$$

where the parameter λ is the expected number of points per sample area. We define x as a random variable whose values are the number of points per sample area. The value of P for a given x is the expected proportion of all occurrences which are x points per sample area. It is assumed that sample

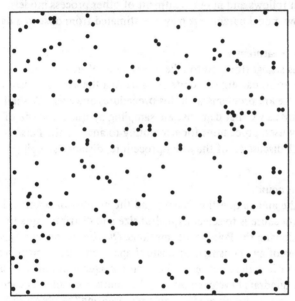

Fig. 2.1 Pattern produced by Poisson process ($n = 200$)

areas are the same size. Fig. 2.1 shows a typical pattern created by Poisson model assumptions. It is interesting to note that

$$P(0;\lambda) = e^{-\lambda} \tag{2.2}$$

There are many situations where such a model is useful. The process responsible for the location of things such as human settlements, store-types, plants and animals, and groups of plants and animals may be approximated by the Poisson process model. If the assumptions do not conform to the nature of the phenomena in question, then the model is inappropriate. In effect, whenever one can assume that within a region locations are equally likely to receive points, then a Poisson process model might be appropriate. In some cases it might be necessary to eliminate some part of a region from consideration because the 'equally likely' assumption cannot be met.

Now, if this model is selected for use, the question arises as to its
verifiability given the observed pattern of points. Unfortunately, one cannot unequivocally verify the assumptions, but there may be reason to have
confidence in the model if one can show that the observed pattern of points
supports the hypothesis that the pattern could have developed from a Poisson
process.

The more knowledge we have of the point process, presumably the better
able we are at selecting a λ value *a priori*. Most often, however, we are not
in a position to assume parameter values before we look at our data. In the
discussion which follows and in our treatment of other process models,
we will show how model parameters may be estimated from observed data.

2.2.   Pattern measurement

It seems straightforward to take a sample of N areas of size a and
compare the observed pattern of points per area to a theoretical Poisson
distribution. There are problems with this procedure, however. We shall take
up these problems as we first demonstrate sampling by quadrats, one of the
two most widely used procedures for attempting to analyze the Poisson
process model. A discussion of the other procedure, distance sampling,
follows.

2.2.1.   Quadrat counts

When the area in question is large and the number of points is large,
say 1000, the procedure is to select a quadrat size and distribute quadrats on
the map area following the Poisson assumptions. *Quadrat* is the name given
to a sampling area of any consistent size and shape — circular, rectangular,
square, etc. More often than not researchers choose square or circular quadrats. After these quadrats have been placed, the points contained in each are
counted and the results are placed in a table representing the observed frequencies (see fig. 2.2 and table 2.1). For the moment disregard the fact that
some points are filled and some are empty. Note that quadrats truncated by
the boundary are discounted.

For many problems the number of points in the area in question is much
smaller than 1000. The procedure in these cases has been to use a lattice
of square quadrats that is placed within the study area. This procedure does
not violate the assumptions of a Poisson process, since we still have independence and each cell is equally likely to be the recipient of a point. The
lattice procedure has the advantage that all points are counted and hence
no information is lost (see fig. 2.3).

Whichever method is adopted the following procedure is used to test the
Poisson model.

(1)   Decide on a quadrat size (see below for a discussion on quadrat
size determination).

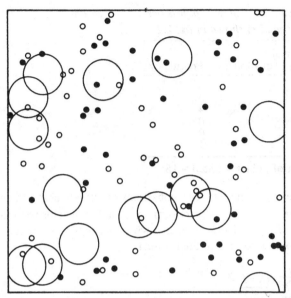

Fig. 2.2 One hundred points and twelve circular quadrats placed by a Poisson process

(2) Determine the mean number of points per quadrat, $\hat{\lambda}$, and consider it to be an accurate estimate of λ.

(3) Find values of $P(x;\hat{\lambda})$ for $x = 0, 1, 2, \ldots$.

In fig. 2.3, $\hat{\lambda} = 2.00$ and the estimated Poisson proportions are

$$P(0;2) = \frac{e^{-2}(2)^0}{0!} = 0.135$$

$$P(1;2) = \frac{e^{-2}(2)^1}{1!} = 0.271$$

$$P(2;2) = \frac{e^{-2}(2)^2}{2!} = 0.271$$

$$P(3;2) = \frac{e^{-2}(2)^3}{3!} = 0.180$$

$$P(4;2) = \frac{e^{-2}(2)^4}{4!} = 0.090$$

$$P(5;2) = \frac{e^{-2}(2)^5}{5!} = 0.036$$

$$P(>5;2) = 1 - [P(0) + P(1) + P(2) + P(3) + P(4) + P(5)] = 0.017$$

Table 2.1. *Frequency of points per circular quadrat shown in fig. 2.2*

Number of points per quadrat	Frequency
0	2
1	5
2	4
3	0
4	0
5	1

Number of points counted = 18

In most cases the applicable test for comparisons of discrete observed and expected frequencies is the chi-square (χ^2) one-sample test. This test requires that the observed frequencies be compared with the frequencies from the hypothesized model according to the formula

$$X^2 = \sum_{i=1}^{k} \frac{(O_i - E_i)^2}{E_i} \tag{2.3}$$

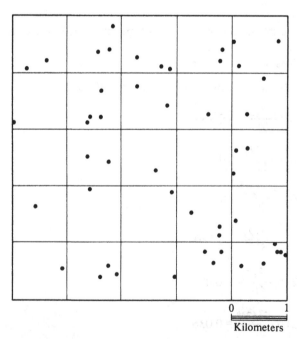

0   1

Kilometers

Fig. 2.3 Fifty points placed by a Poisson process in an area divided into a lattice of twenty-five square quadrats

where O_i is the observed number of cases in the ith category and E_i is the expected number of cases in the ith category of the theoretical model. The number of comparisons made in a test (k) is equal to the number of discrete values of the random variable that have an expected value equal to or greater than 5. If some values of the random variable have less than 5 associated with them, preceding or succeeding values are combined until 5 is reached or exceeded. In order to compare the calculated X^2 with the value of χ^2 shown in statistical tables one must determine the number of *degrees of freedom* (df). The number of df is equal to the number of comparisons made minus one $(k - 1)$. An additional df is lost for each parameter needed to obtain the expected frequencies. Thus, for comparison with the Poisson frequencies, the number of df is equal to the number of comparisons minus two — the parameter λ is needed to calculate the expected frequencies. Calculated values which fall below the χ^2 tabled figures for a predetermined significance criterion indicate that the two sets of frequencies are statistically in accordance. In these cases one might conclude that the observed frequencies may have resulted from a process similar to the one responsible for the development of the theoretical model.*

In this and succeeding tests we will report results by listing the table value for the 80th percentile of the theoretical χ^2 distribution and the calculated X^2 value. This implies that a significance level of 0.20 is the criterion for acceptance or rejection of the null hypothesis of similar frequencies.

Table 2.2, based on fig. 2.3, summarizes the information used for the test. N is the total number of observations (quadrats). The six categories representing the number of points per quadrat were reduced to three in order to satisfy the rule of 5. The tabled χ^2 value of 1.642 indicates that we cannot reject the null hypothesis of no difference between the observed and the expected frequencies and thus we conclude that the two sets of frequencies may have resulted from the operation of similar processes.

Unfortunately, the quadrat size chosen affects the outcome of tests of significance. It is possible for non-Poisson processes to appear as Poisson processes if quadrat size is not chosen judiciously. For example, suppose that a point pattern is the realization of a non-Poisson process; one that causes points to cluster into groups. If the sample quadrats are very small, so that they are only large enough to contain at most one point, then the probability distribution will be binomial with x equal to either 0 or 1. The Poisson distribution, however, can be considered an approximation of the

---

* The X^2 goodness of fit test is used in this book as a guide to our modeling effort. In the next chapters, for simplicity of explanation, the method of developing expected frequencies will not necessarily be carried out using fully efficient methods. In those cases (where maximum likelihood estimates are not used) the use of the X^2 test is not strictly appropriate, but adequate for our purpose. (See Freund, 1971, for an introduction to the various techniques for estimating parameters.)

Table 2.2. *Poisson process model: observed frequency of points taken from fig. 2.3, and expected values*

Number of points per quadrat	Observed frequency	Expected frequency	Expected proportion
0	11 {2	10.15 {3.375	0.135
1	{9	{6.775	0.271
2	5	6.775	0.271
3	{7	{4.500	0.180
4	9 {1	8.075 {2.250	0.090
5	{0	{0.900	0.036
≥6	{1	{0.425	0.017
(N)	25	25	1.000

$X^2 = 0.642$ $df = 1$ $\chi^2_{0.20} = 1.642$

binomial distribution. Thus, a clustered pattern will appear to have been created by a Poisson process.

The first step in quadrat sampling is to find a size which minimizes the variance of the observed set of frequencies. If a point pattern is accentuated by a series of clumps of points, it is possible to create any of a number of different sets of frequencies depending on the way the quadrats are placed on the map and on the quadrat size. There is the tendency for large quadrats to yield high variances; many points in some quadrats and few in others. Large variance may in fact be characteristic of the point pattern, but one should not overemphasize this feature by choosing a large quadrat size. Thus, it is necessary to reduce the variance to a meaningful level by reducing quadrat size. It was just mentioned, however, that very small quadrats can reduce a clustered pattern to a set of frequencies which could have been created by a Poisson process, therefore a small quadrat size is not desirable either. In addition, boundaries of small quadrats will more often cut through a clump of points than those of large quadrats and it is the clumping nature of the pattern which helps us to reject the Poisson process model. Thus some compromise must be found between these extremes. Greig-Smith (1964) suggests that the observed set of frequencies should be reasonably symmetrical. That is, the mean number of points per quadrat should be large enough to allow about as many quadrats with fewer points per quadrat than the mean as there are quadrats with more points per quadrat than the mean. In practice this may be interpreted as selecting a size which will result in observing as many quadrats with no points as quadrats with one or more points. Following this rule of thumb *the appropriate size of quadrat can be approximated as twice the size of the mean area per point.* This seems a reasonable choice in most situations. One must be careful, however, to consider the problems mentioned above before a final choice is made. It is not unusual to select a number of quadrat sizes and repeat the tests, being careful not to infer too much since the tests may correlate with each other. Lattices are helpful in this regard because quadrats can be combined easily. One should also give thought to the rotation of lattices to see if results differ.

Beside Greig-Smith, quadrat size problems have been discussed by Thompson (1958) and Kemp and Kemp (1956). In specific instances, e.g. for discrimination between certain process models, see Pielou (1957) and Rogers and Gomar (1969). In geography quadrats are used for many problems. These will be considered when particular spatial models are discussed.

There are several other tests besides the X^2 that may be useful in deciding on the validity of the Poisson process model. The simplest test is based on the property of a Poisson distribution that the variance about the mean is equal to the mean. A comparison of these two values, v and m, obtained from the observed set of frequencies can be evaluated using a t test with standard error $\sqrt{[2/(N-1)]}$ and $N-1$ degrees of freedom, where N is the number of observations (quadrats).

$$t = \frac{v - m}{\sqrt{[2/(N-1)]}} \qquad (2.4)$$

Using the data from table 2.2 we find that the mean number of points per quadrat is 2.00, the variance is 1.76, and t is -0.83, which is well within the curve of the normal probability distribution, indicating no significant difference between the two values.

Although not designed as an index of 'Poissonness', researchers find that the t value provides information about the degree of departure from Poisson expectation. A large positive t value indicates a departure in the direction of high variance or clumping (clustering), while negative t values indicate low variance or evenness. If either of these outcomes are sufficient to have the hypothesis of the Poisson model rejected, then there is some information as to the kind of model that is needed to explain the pattern. Pielou (1969) reviews much of the work on the variance—mean ratio.

Taking this further, an alternate test called the *index of dispersion* tells us the degree to which a pattern is not Poisson-like in the direction of greater dispersion or more clustering. The index of dispersion (Pielou, 1969, p. 91) is given as

$$\sum_{i=1}^{N} \frac{(x_i - m)^2}{m} \qquad (2.5)$$

which is the sum of N terms (number of quadrats). The value x_i represents the number of points contained within the ith quadrat and m is the mean number of points per quadrat. Since x_i and m can be viewed as observed and expected values respectively, the sum is therefore approximately distributed as χ^2 with $N-1$ degrees of freedom. The probability of obtaining any value of the index of dispersion may be tested by way of a χ^2 table. Index values greater than the tabled χ^2 value indicate a tendency toward clustering while values less than the tabled χ^2 value imply a tendency toward dispersion. Quadrats must be grouped in order to allow the expected values to be equal to at least 5.

2.2.2. Distance measures

In order to avoid quadrat sampling problems, one can make tests on measurements of distance between sampled points and their nearest neighboring points. In quadrat analysis, two sets of frequencies are compared, the observed and the expected, while in the distance approach the observed mean distance taken from the nearest neighbor measurements and the expected mean distance are compared. In addition, moments about the mean and mean distances to second, third, etc., nearest neighbors can be used for comparison. The expected mean distance is derived from our knowledge of the way the theoretical points would be placed on a map, i.e. in the manner of the Poisson process assumptions.

Recall the Poisson probability distribution shown in (2.1). Now let the specified area under consideration be a circle of radius r so that $a = \pi r^2$. The value $e^{-\lambda}$ is the probability that a randomly chosen area of size a will contain no points. This is the same as the proportion of distances to nearest neighbor greater than or equal to r. The mean of r, called $E(r)$, is the expected mean distance to nearest neighbor in a Poisson process generated pattern. Clark and Evans (1954), show this to be

$$E(r) = \frac{1}{2\sqrt{\lambda}} \tag{2.6}$$

where λ is the density (points per unit area). Note that previously λ represented the expected number of points per quadrat of size a. The standard error of $E(r)$ is

$$\sigma_r = \frac{0.26136}{\sqrt{N\lambda}} \tag{2.7}$$

where N is the number of measurements made. The significance of the departure from the Poisson process model can be assessed using

$$Z = \frac{\bar{r} - E(r)}{\sigma_r} \tag{2.8}$$

where Z is the standard variate of the normal curve and $\bar{r}$ is the observed mean distance to nearest neighbor. In order to calculate λ, one must count the total number of points in the region (n) and not just the sampled points (N). Equation (2.8) is particularly applicable for samples of 100 or more.

This approach is more tedious than the quadrat method but its advantages are that there is no quadrat size problem and more observations are possible (the number of map points usually exceeds the number of quadrats). In addition, distance measures take account of the distances separating points which the quadrat method disregards. A further advantage of the distance approach is that when testing a Poisson process model one can use either distances from sampled points to their nearest neighboring points (called point-to-point distances, r_{pp}) or distances from sampled locations (chosen by allowing all coordinate locations to be equally available for selection) to their nearest neighboring points (called location-to-point distances, r_{lp}). With point-to-point distances one either uses each point or a sample of points. The sampling procedure requires that all points be labeled so that each has an identical probability of being selected. This is a tedious procedure, especially when the total number of points is large. Location-to-point measurements alleviate the problem. In both cases, however, one must count (or estimate) the total number of points because the density measure depends on it.

The test for significance is based on differences between r_{pp} and r_{lp} (Pielou, 1969, pp. 115–17). In a Poisson process generated pattern the means and variances of these two sets of measurements are equal. Thus, differences indicate a non-Poisson process. Here we take N distances of each kind and square and sum them. The ratio of the two is called C.

$$C = \frac{\Sigma r_{lp}^2}{\Sigma r_{pp}^2} \qquad (2.9)$$

The expected value of C is 1. The test proposed by Pielou (1969) is based on the standardized normal variate

$$Z = 2(y - \tfrac{1}{2})\sqrt{(2N + 1)} \qquad (2.10)$$

for $N > 50$ where $y = C/(1 + C)$. The expected value of y is $\tfrac{1}{2}$. This test is particularly accurate for $N > 50$.

There are a number of problems in dealing with distance measurements. First, since it is necessary to know the density (λ) of the pattern, it becomes imperative that the area under study conforms exactly to the Poisson assumption of equally likely locations for points. (In chapter 6, a method for measuring pattern in a density-free setting is discussed.) Consider the case when a small cluster of points is included in a large area. The density is low. For the case when the same cluster of points is in a small area, the density will be considerably higher. The results of nearest neighbor analyses in the two cases will be different. To avoid this problem it is usually necessary for the sample area to be positioned within the total pattern of points (see fig. 2.4).

Second, measurements to nearest neighbors from points close to the boundary of the study area may be biased if the nearest neighbor distance is greater than the distance of the point to the boundary. In a Poisson process

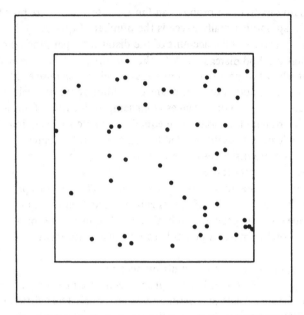

Fig. 2.4 Only in the inner bounded area is there confidence that the Poisson process model holds. Note that the density of points would differ considerably depending on whether the inner or outer boundary is used

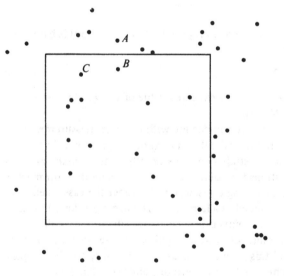

Fig. 2.5 Since B is closer to A than C the measurement BA should be included in the sample, but not AB

we might expect to find a point outside of the study area boundary that is closer to the point in question than the point is to its nearest neighbor within the study area. Thus, as a practice, measurements should be made from sample points or locations within the study area to nearest neighbors whether or not they are contained within the study area (see fig. 2.5). An alternative procedure would be to eliminate from the sample measurements from those points closer to the boundary than they are to their within-area nearest neighbor. This procedure would be necessary if only the study area and not the area surrounding it satisfied the Poisson assumption (see fig. 2.6).

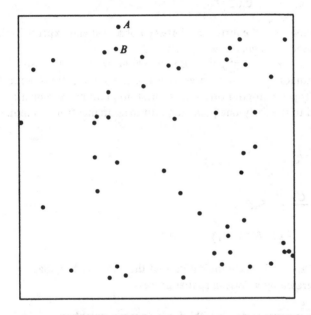

Fig. 2.6 No measurement would be taken from point *A* since it is closer to the boundary of the study area than it is to its nearest neighbor *B*

Third, a problem in interpretation may result when points are in small clusters but the clusters are long distances from one another. The mean distance to nearest neighbor will be short and thus the hypothesis of a Poisson generated process would be rejected. What may confuse us is the fact that the pattern of clusters and not the pattern of points is the realization of a Poisson process. If it is the clusters that are the subject of a Poisson process model then the clusters must be represented as points.

Forty-one nearest neighbor measurements were made on fig. 2.3 following the procedure of eliminating a measurement if the sample point is closer to the boundary than to its nearest neighbor. The results are

$$E(r) = \frac{1}{2\sqrt{\hat{\lambda}}} = 0.3536 \text{ kilometers}$$

where $\hat{\lambda}$ equals 2,

$$\sigma_r = 0.26136/\sqrt{N\hat{\lambda}} = 0.0289$$

$$\bar{r} = 0.3061 \text{ kilometers}$$

and $$Z = \frac{\bar{r} - E(r)}{\sigma_r} = -\frac{0.0475}{0.0289} = -1.64$$

These results imply that the difference between observed and expected values does not vary more than chance would have it.

Fig. 2.2 has the same set of points as is shown in fig. 2.3 but it also includes fifty randomly selected places used as centers of location-to-point measurements (open points). Forty-one location-to-point measurements were compared to the forty-one point-to-point measurements just summarized. The results are

$$C = \frac{\Sigma r_{lp}^2}{\Sigma r_{pp}^2} = \frac{5.915}{4.940} = 1.197$$

$$y = \frac{C}{1+C} = 0.545$$

$$Z = 2\left[y - \tfrac{1}{2}\right]\sqrt{(2N+1)} = +0.818$$

Differences are small, again indicating that the point pattern may well have been generated by a Poisson spatial process.

2.2.3.   Measurements to second, third, etc., nearest neighbors

It is sometimes helpful to further verify the existence of a Poisson generated pattern by measuring distances to neighbors other than the first nearest neighbor. If there are special nuances characteristic of the observed distribution which are non-Poisson in nature, such as a tendency for doublets or triplets of points to cluster, further testing may be necessary. A method has been devised which allows for the comparison of observed distances to the second, third, etc., nearest neighbors with the distances expected as a result of a Poisson model (Thompson, 1956).

The expected mean distance to the jth nearest neighbor is derived from the Poisson distribution. This value is

$$E(r_j) = \frac{1}{\sqrt{\lambda}} \frac{(2j)! j}{(2^j j!)^2} \qquad (2.11)$$

where λ is the expected number of points per unit area. Table 2.3 gives the expected mean distance $E(r_j)$ and standard deviation for the first five nearest neighbors.

Table 2.3. *Expected mean distance and standard deviation for nearest neighbors 1−5 in a Poisson process generated spatial pattern*

Nearest neighbor j	Expected mean distance $E(r_j)$ $\times 1/\sqrt{\lambda}$	Expected standard deviation σ_{rj} $\times 1/\sqrt{\lambda}$
1	0.5000	0.2614
2	0.7500	0.2723
3	0.9375	0.2757
4	1.0937	0.2774
5	1.2305	0.2782

The normal deviate

$$Z = \frac{\bar{r}_j - E(r_j)}{\sigma_{rj}/\sqrt{N}} \tag{2.12}$$

may be used as a test against the probability levels of the normal distribution. $\bar{r}_j$ is the mean of N observed values of nearest neighbor measurements.

The results of five tests (table 2.4) on data taken from fig. 2.3 indicate that except for second nearest neighbor distances the pattern does not deviate considerably from Poisson expectations. It does seem, however, that since all Z values are negative a grouping tendency may be the case. Note that due to the boundary rule mentioned in the previous section, fewer measurements are possible as nearest neighbor distances increase, i.e. as one measures distances to successive nearest neighbors.

Cowie (1968) and Roder (1974, 1975) suggest that it is helpful to examine the complete frequency curve for each set of j-neighbor distance measurements. These frequencies can be compared to theoretically derived frequencies by means of either a chi-square test or a Kolmogorov−Smirnov one-sample

Table 2.4. *Observed and expected mean distances to first five nearest neighbors for the pattern shown in fig. 2.3 (in kilometers)*

Nearest neighbor	Number of measurements	Expected mean distance	Observed mean distance	Expected standard deviation	Z
1	41	0.3536	0.3061	0.1848	−1.64
2	38	0.5303	0.4353	0.1925	−3.04
3	33	0.6629	0.6003	0.1949	−1.85
4	25	0.7734	0.7060	0.1962	−1.72
5	23	0.8701	0.8013	0.1967	−1.68

test. (See p. 97) Clark and Evans (1954) and Dacey (1962) discuss what has been called a regional method of measurement to the first six nearest neighbors of a point. Most researchers now agree that the regional method is an unnecessary complication, fraught with difficulties, therefore it will not be discussed.

2.2.4. Reciprocal point pattern measurements

When two points are the nearest neighbors of each other the relationship is said to be reciprocal or reflexive. The two points are closer to each other than either is to any other point. For a Poisson process model, the expected number of reciprocal points can be calculated. (See Clark and Evans, 1955; Clark, 1956; Dacey, 1969a.) The probability that a point A will form a reciprocal pair with B is

$$\frac{6\pi}{8\pi + 3\sqrt{3}} = 0.6215 \qquad (2.13)$$

The right hand column of table 2.5 shows the first five theoretical values taken from Dacey (1969a). As would be expected the proportions fall off as j increases. In table 2.5 we also give the observed proportion of reciprocal pairs from fig. 2.3 up to the fifth nearest neighbor (each reciprocal pair counts two since two observations are made).

Table 2.5. *Observed and expected proportion of reciprocal pairs, first to fifth nearest neighbors (based on fig. 2.3)*

Nearest neighbor	Measurements	Observed proportion of reciprocal pairs	Expected proportion of reciprocal pairs
1	41	0.5366	0.6215
2	38	0.2105	0.3291
3	33	0.2424	0.2430
4	25	0.1600	0.2016
5	23	0.1739	0.1760

Judging from our earlier results and the fact that no test of significance is available for reciprocal pairs, it must be concluded that this is a very crude measure of pattern. It can however roughly indicate if some systematic characteristic is embedded in the observed pattern. For example, in table 2.5 the observed values for first and second nearest neighbors are less than the expected values, indicating that a tendency exists for points to be slightly more dispersed than would be expected under the Poisson rules. Our earlier results for second nearest neighbor distances indicate just the opposite. Large samples (hundreds of observations) are needed for this approach to be feasible. But then, this approach might be more quickly applied when the number of measurements is large. A further useful characteristic of this approach is that the measure is density-free, i.e. no λ value is needed.

Dacey (1960*b*) demonstrates a one-dimensional approach to reciprocal pairs by studying the spacing of river towns (points along a line). Porter (1960) found a problem in dealing with the measurements between first nearest neighbor towns and was critical of the approach. Perhaps a test, if one can be developed, should include information about first, second and further nearest neighbors (see section 6.3.2 on Delaunay triangles).

The literature on nearest neighbor and similar distance measures is extensive. Some researchers instrumental in its development are Skellam (1953), Morisita (1954), Moore (1954), Pielou (1959, 1962), Holgate (1965), Thompson (1956), Dacey (1960*a*, 1963*a*, 1964*a*), Blacklith (1958), and Clark and Evans (1954, 1955). In geography, the work of Clark and Evans and the various modifications and extensions by Dacey have stimulated much empiricism. Most noteworthy are the studies by King (1961, 1962) on urban settlement spacing, Getis (1964) and Clark (1969) on store location patterns, Hodder (1972*b*) on the spacing of Romano-British walled towns, and Dacey (1962) on the spacing of central places. The boundary problem has been discussed by King (1969) and a useful measurement modification to account for the effect of boundaries has been provided by Hsu and Mason (1974). Norcliffe (1967) has developed a nearest neighbor technique for grouping points, and Sorensen (1974) discusses a method for measuring the spatial association between point patterns.

Criticism has been leveled at the various uses and techniques of nearest neighbor analysis. Much of the work mentioned above was stimulated by the problems which arise in dealing with distance measures. For example Pielou (1962) discusses location-to-point measures as a reaction to the tediousness of measuring point-to-point distances. In biology there is concern for estimating the density of plants or animals in a given area. Often a sample of nearest neighbor distances are used to estimate the density, but Persson (1971) and Holgate (1972) show that if the pattern of points is clustered or regular, estimates of density may be seriously inaccurate. Review articles by Pinder and Witherick (1972), Susling (1971) and De Vos (1973) summarize the state of nearest neighbor analysis in geography and consider problems of interpretation.

There are many technical treatises on the nature of the Poisson distribution. Most of this work is summarized in Johnson and Kotz (1969) while Feller (1968) and Miles (1970) discuss the spatial or planar Poisson point process. An analysis of the Poisson distribution is contained in Haight (1967). No comprehensive geographical treatment of Poisson-process models would be complete without mention of the classical description of the location of villages in the Tonami Plain of Japan. Matui (1932) completed this work long before other geographers used probability models for pattern study. The paper is reprinted in Berry and Marble (1968).

2.2.5. Segregation and symmetry

The idea of nearest neighbor distances can be used to identify segregation in a two-species population. Let us consider this problem briefly. Suppose that the m members of one species are each called A and the n of the other B. The frequency of A's as centers of measurement where B's are their nearest neighbors can be entered into cell b in a two by two table (see below). Cell a contains the frequency of A's that are nearest neighbors of A's and cells c and d contain the frequency of nearest neighbors measured from B. If both the pattern of A's and of B's resulted from a Poisson process in nature we would expect the cells to contain values in proportion to the number of points in each pattern. For the case when $m = n$ the expectation is $a = b = c = d$.

		Nearest neighbor		
		A	B	
Measurement from	A	a	b	m
	B	c	d	n
		r	s	N

A χ^2 criterion may be used to judge whether observed cell frequencies differ significantly from expectation. For cell a, the expected value is mr/N, for b the expected value is ms/N, for c, nr/N and for d, ns/N. If b and c are greater than their expected values, the species are spatially related. If, however, their frequency is less than expected then the species are at least partly segregated.

Pielou (1961) has suggested that a coefficient of segregation be defined as

$$S = 1 - \frac{\text{observed number of } b \text{ and } c \text{ pairs}}{\text{expected number of } b \text{ and } c \text{ pairs}}$$

which can be written

$$S = 1 - \frac{N(b + c)}{ms + nr} \tag{2.14}$$

For a perfectly segregated pattern, $b = c = 0$ and $S = +1$. When the patterns of A's and B's are both realizations of a Poisson process then S is close to zero. If B's are spatially associated with A's then S will be between 0 and -1, and exactly -1 in the case when $m = n$.

From fig. 2.2 we identify two species, one represented by filled points (A) and the other by empty points (B). The frequency of each type of association is summarized in the table below.

		Nearest neighbor		
		A	B	
Measurement	A	24	19	43
from	B	22	22	44
		46	41	87

Since 13 measurements failed the boundary test, $N = 87$. Thus

$$S = 1 - \frac{87\,(19 + 22)}{(43)\,(41) + (44)\,(46)} = +0.058$$

which is very close to a Poisson process expectation.

In the social science and ecological literature measures of segregation are usually based on the occurrence of one, both or none of each species within designated areas. Pielou (1969, pp. 159–71) treats this subject. Contiguity measures have been used to indicate when areas (not points) designated as A and B are spatially related (Cliff and Ord, 1973). Pielou (1969, pp. 179–86) considers segregation among many species in an area by using a transect through the area. Species are identified and labeled on a line representing the transect. The sequence is then studied following various probabilistic approaches.

2.3.   Truncated Poisson process model

As an alternative to the quadrat study of point patterns one can consider the number of points which are members of groups. In this section we discuss how the size frequencies of groups of points can be compared to Poisson model expectations. A group is considered to contain one or more members. The hypotheses that give rise to tests are aspatial, i.e. group size not group location is of central concern. Later we will add a spatial dimension to studies of group sizes, but for the moment let us consider individuals assigned to possible groups according to the assumptions of equal probability and independence. Under these assumptions group sizes will be Poisson distributed but without the category 0 since a group, to be observed, must have at least one member. Thus we develop a Poisson model with the impossible outcome, 0, truncated from the distribution.

The distribution is derived by first subtracting $e^{-\lambda}$ from the Poisson model since this represents the 0 category and then allocating this amount proportionately to all categories greater than 0. The result is

$$P(x; \phi) = \frac{\phi^x}{(e^\phi - 1)x!} \quad \text{for } x = 1, 2, \ldots \tag{2.15}$$

where $P(x; \phi)$ stands for the expected proportion of all group sizes which are of size x. Each possible value of ϕ will result in a different truncated Poisson distribution. The theoretical mean is

$$\mu = \phi/(1 - e^{-\phi}) \tag{2.16}$$

It is necessary to estimate ϕ from the observed frequencies. Cohen (1960) provides a nomograph using the observed mean (m) as an estimate of μ. Irwin (1959) derives an estimate ($\hat{\phi}$) of the parameter by the following expression:

$$\hat{\phi} = m - \sum_{j=1}^{\infty} \frac{j^{j-1}}{j!} (me^{-m})^j \qquad (2.17)$$

where j takes values 1, 2, 3, . . . until the sum is not appreciably changed by the addition of further terms. In most instances, especially with $m > 2.5$, j need only be taken to 3 or 4 for reasonably accurate estimates of ϕ.

Once ϕ is estimated, the various values of the probability distribution can be calculated. In table 2.6 the data from one of many experiments performed

Table 2.6. *Truncated Poisson process model: group size in free play of four-year-olds at Cambridge, Mass., nursery school*

Group size	Observed frequency	Expected frequency
1	385	376.8
2	123	125.8
3	25	28.0
≥ 4	3	5.4
(N)	536	536

$X^2 = 1.629 \qquad df = 2 \qquad \chi^2_{0.20} = 3.219$

Source: Cohen (1971, p. 84).

by Cohen (1971, p. 84) are given. He observed the group size in free play of four-year-olds in a nursery school in Cambridge, Massachusetts. His theory involved the tendency of humans and animals to group into certain size frequencies. He was not concerned with the location of the groups relative to each other. In table 2.6 the mean group size is 1.34 and ϕ was estimated (by computer for $j = 10$) as 0.668. The X^2 results indicate that the observed and expected frequencies are not sufficiently different to reject the hypothesis of a truncated Poisson model.

In summary, then, the assumptions needed for specifying a truncated Poisson distribution (truncated at 0) are:

(1)   each possible group is equally likely to receive a member;
(2)   each member is placed in a group independently of the placement of any other member in any other group;
(3)   possible groups receiving one or more members are of consequence.

A discussion of the truncated Poisson distribution and review of the technical literature is given in Johnson and Kotz (1969). Cohen (1971) describes a series of studies where use is made of the model. Most prominent is the work of James (1951, 1953) published in full by Coleman (1964). James found that the truncated Poisson model closely described free-forming groups, i.e. groups whose members are relatively free to maintain or break off contact with one another. He also observed that the truncated Poisson model describes pedestrian and shopping groupings, play groups, and swimming pool gatherings. In a prelude to a study of group location Getis and Merk (1973) found that the frequency of group sizes of people on beaches could be considered the realization of a truncated Poisson process.

Several other non-Poisson distributions have been used as models of group size. Most popular are the truncated negative binomial distribution and the logarithmic series distribution. The assumptions which define these models are different than those for the truncated Poisson model. More will be said about these in the next chapter.

3
Point patterns: mixed Poisson process models

The Poisson process model discussed in chapter 2 is clearly inadequate for many situations. In chapter 1 we mentioned such processes as agglomeration, diffusion, birth and death which very often do not conform to the two rather limiting assumptions of the Poisson process model. The class of models particularly well suited for describing more complex spatial processes is based on mixing the Poisson process rules with the assumptions inherent in other distributions. One can go further, of course, and mix combinations of random variables all of which are not associated with the Poisson probability function, but, at present, there seems to be little need to do so.

Mixed Poisson process models may be divided into two general types: *additive* or *multiplicative*. Additive processes are simply the sum of several independent groups of assumptions. Multiplicative processes are more complex. They are developed from a mixture of groups of assumptions. Sometimes one group of assumptions is conditioned by another (*generalized models*) and sometimes the mixtures are a result of joining independent sets of assumptions (*compound models*).

3.1. Additive processes

There are several kinds of mixed Poisson process models which can be developed to describe situations where several location processes are acting in an additive way. One kind of model is based on two or more independent spatial processes which act in the same region. These are called *superimposition* models. The other model depends on two or more independent processes each of which acts in separate but spatially contiguous areas. These are called *heterogeneous* models since the implication is that the separate areas contain dissimilar patterns. Map patterns produced by additive processes are shown in fig. 3.1.

3.1.1. Superimposition

When it is suspected that two or more patterns are superimposed one on another the resulting pattern can be modeled by adding independent

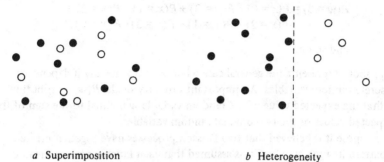

a Superimposition   *b* Heterogeneity

Fig. 3.1 Patterns resulting from additive processes. (*a*) Superimposition.
(*b*) Heterogeneity (the dashed line represents the subarea boundary)

discrete random variables. Suppose that each of two patterns is the realization of a process represented by the random variables x and y, respectively.
The pattern produced by the two processes taken together is represented
by the random variable w. The diagram shows how values of w are related

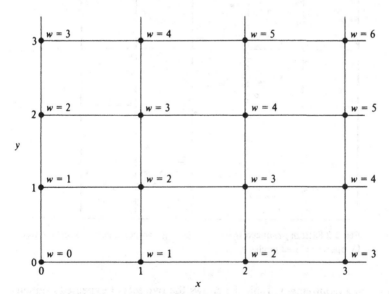

to values of x and y. The probabilities associated with the values of the new
random variable w are obtained as follows

$$P(w = 0) = P(x = 0) \cdot P(y = 0)$$

$$P(w = 1) = P(x = 0) \cdot P(y = 1) + P(x = 1) \cdot P(y = 0)$$

$$P(w = 2) = P(x = 0) \cdot P(y = 2) + P(x = 1) \cdot P(y = 1) +$$
$$P(x = 2) \cdot P(y = 0)$$

$$P(w = 3) = P(x = 0) \cdot P(y = 3) + P(x = 1) \cdot P(y = 2) +$$
$$P(x = 2) \cdot P(y = 1) + P(x = 3) \cdot P(y = 0)$$

and so on.

The above represents the general case where x and y are any independent, discrete random variables. An important property of the $P(w)$ distribution is that the expected value of the random variable w is equal to the sum of the expected values of the two original random variables.

Suppose it is believed that two Poisson processes have together created a pattern in a region where it is assumed that each Poisson process created part of the pattern. The pattern shown in fig. 3.2 represents the two independent sets of objects. In this case each has the same density, although

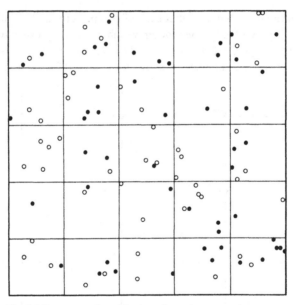

Fig. 3.2 Pattern produced by superimposing two point patterns represented by open or closed circles

this is not a requirement. Table 3.1 shows the two sets of expected frequencies, the expected combined frequencies, and the observed values. The χ^2 value at the 0.20 level is 1.642 for $df = 1$. The observed value of the test is 0.377 which shows that observed and expected frequencies conform rather well.

Poisson plus Poisson process model. The distribution of w may be obtained in another, less tedious way. In the last experiment the mean of the expected

Table 3.1. *The addition of two Poisson random variables: observed and expected values*

Points per quadrat	$P(x; \lambda)$	$P(y; \lambda)$	$P(w)$	Frequency observed	Frequency expected
0	0.135	0.135	0.018	0	0.450 ⎫
1	0.270	0.270	$0.073 = 2(0.036)$	1	1.825 ⎬
2	0.270	0.270	$0.146 = 2(0.036) + 0.073$	5	3.650
3	0.180	0.180	$0.194 = 2(0.024) + 2(0.073)$	7	4.850 ⎫
4	0.090	0.090	$0.194 = 2(0.012) + 2(0.049) + 0.073$	4	4.850 ⎬
5	0.036	0.036	$0.131 = 2(0.005) + 2(0.012) + 2(0.049)$	2	3.275 ⎫
6	0.012	0.012	$0.104 = 2(0.002) + 2(0.010) + 2(0.024) + 0.032$	1	2.600 ⎬
7	0.003	0.003	$0.059 = 2(0.004) + 2(0.003) + 2(0.010) + 2(0.016)$	4	1.475 ⎬
$\geqslant 8$	0.002	0.002	0.081	1	3.025 ⎭
(P, N)	1.000	1.000	1.000	25	25
$X^2 = 0.377$	$df = 1$	$X^2_{0.20} = 1.642$			

frequencies is equal to 4 since $\lambda = 2$ in each case. We can write $\lambda_w = \lambda_x + \lambda_y$ and

$$P(x;\lambda_w) = \frac{e^{-\lambda_w}\lambda_w^{\,x}}{x!} \quad \text{for } x = 0, 1, 2, \ldots \tag{3.1}$$

which is itself a Poisson distribution. We shall name the distribution 'Poisson plus Poisson'. Adding (3.1) to another Poisson random variable gives us another Poisson plus Poisson distribution.

Since by adding two Poisson process generated random variables one gets a new Poisson distribution, it is clear that verifying an hypothesis of the existence of a Poisson process does not necessarily tell us that two (or more) independent Poisson processes are responsible for the pattern. If any test based on (3.1) is to be meaningful, it must be clear at the start that several Poisson processes are responsible for the pattern.

In summary, then, the assumptions for a Poisson plus Poisson process model are that:

(1)   more than one location process exists in the region under study.
(2)   for each process each location is equally likely to receive a point.
(3)   each assignment of a point is independent of all other assignments of points.

Poisson plus Bernoulli process model. If a random variable associated with a Poisson process were added to a non-Poisson random variable a non-Poisson probability function would be needed to describe the resulting probability distribution. One of the simplest superimposed models of this type is developed by adding a Poisson random variable to a Bernoulli random variable. The Bernoulli distribution is a special case of the binomial distribution. Instead of n trials there is just one trial which can result in a success (1) or a failure (0).

Recall that the binomial is

$$\binom{n}{x}\theta^x(1-\theta)^{n-x}$$

where θ is the probability of success and $(1-\theta)$ is the probability of failure. Setting $n = 1$ we get

$$\theta^x(1-\theta)^{1-x} \quad \text{for } x = 0, 1 \tag{3.2}$$

which is the Bernoulli distribution. (Note that both $\binom{1}{0}$ and $\binom{1}{1}$ are equal to one.)

The question that arises is, what kind of point process would demand this mixture of random variables? Suppose it was suspected that a pattern created by a Poisson process was being disturbed by some systematic point process which tended to add a point to each subarea of the region under consideration. In studies of central places it may be thought that a non-Poisson evening process is at work in addition to a Poisson process. Dacey (1964*b*, 1966*b*), for example, has suggested that the existence of county seats in Iowa helped to bring about a more even pattern of towns than one generated by a Poisson process. In chapter 4 several other approaches to the study of even patterns are discussed.

By joining the Bernoulli random variable to the spatial Poisson random variable we are, in effect, adding one point to each of the subareas with probability θ, or not adding one point to each of the subareas with probability $(1 - \theta)$. If x represents the Poisson random variable ($x = 0, 1, 2, 3, \ldots$) and y the Bernoulli random variable ($y = 0, 1$) then w could be enumerated following the procedure shown above. The diagram indicates how the values of the

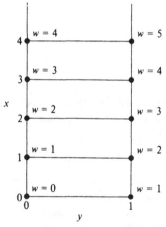

random variables are related. The values of $P(w)$ can be found by

$$P(w = 0) = P(x = 0) \cdot P(y = 0)$$

$$P(w = 1) = P(x = 1) \cdot P(y = 0) + P(x = 0) \cdot P(y = 1)$$

$$P(w = 2) = P(x = 2) \cdot P(y = 0) + P(x = 1) \cdot P(y = 1)$$

$$P(w = 3) = P(x = 3) \cdot P(y = 0) + P(x = 2) \cdot P(y = 1)$$

and so on.

Since the expected result of a Bernoulli trial is either θ or $(1 - \theta)$ depending on the expected existence or non-existence of a success we may join the Poisson expression with θ, then with $(1 - \theta)$, and sum. Thus, we have the single expression

$$P(x; \lambda, \theta) = \frac{\theta e^{-\lambda}\lambda^{x-1}}{(x - 1)!} + (1 - \theta)\frac{e^{-\lambda}\lambda^{x}}{x!} \quad \text{for } x = 0, 1, 2, \ldots \quad (3.3)$$

which is the Poisson plus Bernoulli distribution. As a result of obtaining a
successful trial with probability θ, the value of x in the first term on the
right hand side of the equation must be reduced by one. Since the failure,
with probability $(1 - \theta)$, does not result in an additional point being added
to the pattern, no change in x is needed in the second term on the right hand
side. The two terms are added simply because the model considers the existence *or* non-existence of a successful Bernoulli trial. To make the model
usable (ridding ourselves of $(x - 1)!$ in the denominator) multiply the first
term on the right by x/x which then gives

$$P(x; \lambda, \theta) = \theta x \frac{e^{-\lambda}\lambda^{x-1}}{x!} + (1 - \theta)\frac{e^{-\lambda}\lambda^{x}}{x!} \quad \text{for } x = 0, 1, 2, \ldots \quad (3.4)$$

Dacey (1964b) derives the mean as

$$\mu = \lambda + \theta$$

and the variance

$$\sigma^2 = \lambda + \theta(1 - \theta)$$

Both θ and λ can be estimated using sample data. The theory that is being
tested, however, could suggest to the researcher the value of θ. Dacey
(1964b) has found that $\hat{\lambda}$ (the estimate of λ) is

$$m - \hat{\theta} \quad \text{where} \quad \hat{\theta} = \sqrt{(m - v)}$$

The observed mean is m, and v is the observed variance. Note if the variance
is greater than the mean, indicative of clustering, the model is inoperative.

To the Poisson created pattern of fig. 2.3 points were added according to
the Bernoulli process with $\theta = 0.5$. Thus, each of the twenty-five cells had
a 0.5 chance of receiving an additional point. With the use of a random
numbers table fourteen points were assigned so that the pattern now contained sixty-four points instead of fifty. Table 3.2 gives the resulting frequencies. For the purposes of this test θ and λ were estimated from the
observed frequencies. These are $\hat{\theta} = 0.613$ and $\hat{\lambda} = 1.947$. The mean is 2.56
and the variance 2.184. The estimated θ is larger than the theoretical
Bernoulli θ because by chance more than 12.5 points were allocated to the
25 cells. The X^2 value is 0.372 indicating, as expected, little difference between
the observed and expected frequencies. If $\theta = 0.5$ were used the observed X^2
value would be 0.539.

In summary, the assumptions for a Poisson plus Bernoulli model are:

(1)   all locations in study area are equally likely to receive a point
(2)   the location of each point is independent of the location of any
       other point.

Table 3.2. *Poisson plus Bernoulli process model: observed and expected frequencies*

Points per quadrat	Observed frequency before addition	Observed frequency after addition	Expected frequency
0	2	0	⎰ 1.375
1	9	6	⎱ 4.875
2	5	8	6.875
3	7	6	5.850
4	1	3	⎧ 3.525
5	0	1	⎨ 1.625
6	1	0	⎪ 0.625
≥7	0	1	⎩ 0.250
(N)	25	25	25

$X^2 = 0.372$ $df = 1$ $\chi^2_{0.20} = 1.64$

(3) each subarea (quadrat) is equally likely to receive an additional point. It is optional whether θ is specified *a priori*.

(4) The location of the additional point in a subarea is independent of the location of the additional point in any other subarea.

3.1.2. Heterogeneity

The term 'heterogeneity' is used to describe the class of models which are based on the assumption that one sampling area of a region is more likely to receive a point than is another sampling area. This is a useful assumption to make for the many known situations where probabilities vary according to the nature of the physical and human environment. There are several ways in which one may model heterogeneous processes (sometimes called inhomogeneous processes). One way is to create probability functions which are based on the addition of random variables. Each probability function represents a process operating in one of the sampling areas. Weighting of the functions would be needed to account for areas of different size. A second way is to introduce a probability function which represents the variation in the pattern from sampling area to sampling area. This second kind of model is characteristic of a multiplicative process which will be discussed in section 3.2. The additive models are special cases of multiplicative models.

Double Poisson process model. Suppose that because of some restraint on development in an area the density of the points is lower in one area than in an area contiguous to it. The heterogeneous model describes the pattern in the region as a whole. One cannot use the Poisson plus Poisson process model of the last section as the process responsible for this situation, since that distribution is itself a Poisson distribution. In this case it is evident that the Poisson model assumptions do not hold for the entire region since all locations are not equally likely to receive points. What we need here is the

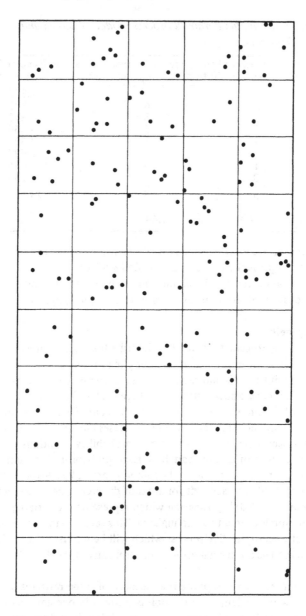

Fig. 3.3 Pattern resulting from heterogeneous point process. Note that the density is higher in the upper 25 quadrats

development of a model which takes into account two spatially separated processes and combines them into one model. If the process in both areas is assumed to be Poisson, the resulting distribution is called a *double Poisson*. The double Poisson is given by

$$P(x; \lambda_1, \lambda_2) = \tfrac{1}{2} \frac{e^{-\lambda_1} \lambda_1^x}{x!} + \tfrac{1}{2} \frac{e^{-\lambda_2} \lambda_2^x}{x!} \quad \text{for } x = 0, 1, 2, \ldots \quad (3.7)$$

Unlike the Poisson plus Poisson distribution, the double Poisson distribution requires two parameters.

In fig. 3.3 there are two point patterns, each capable of being created by a Poisson process. The quadrats are divided into two groups of twenty-five. The pattern in the upper half contains one hundred points and the lower half has fifty points. The points were assigned following the Poisson model assumptions. We shall assume that the distributions are weighted equally, thus we are able to assign the double Poisson coefficients of $\tfrac{1}{2}$ to each term as shown in (3.7). The two parameters λ_1 and λ_2 are estimated by:

$$\left. \begin{aligned} \hat{\lambda}_1 &= m + \sqrt{(v - m)} \\ \text{and} \quad \hat{\lambda}_2 &= m - \sqrt{(v - m)} \end{aligned} \right\} \quad (3.8)$$

where m and v are the mean and variance of the observed frequencies for the entire pattern. In our example the variance is 3.918 and the mean is 3.00, thus $\hat{\lambda}_1 = 3.958$ and $\hat{\lambda}_2 = 2.042$. In a double Poisson model the variance is greater than the mean indicating that there is a tendency for the points to cluster. The clustering is a result of the greater density of points in one of the subareas. For fig. 3.3, table 3.3 gives the result of a test on the double Poisson distribution. As expected, the observed and expected frequencies are similar. The χ^2 criterion at the 0.20 level of significance for $df = 2$ is 3.22 which is well above the observed value of 0.994.

General double Poisson process model. When one cannot assume equally weighted random variables the estimation procedure is more tedious. The distribution for these cases is

$$P(x; \lambda_1, \lambda_2, a_1, a_2) = a_1 \frac{e^{-\lambda_1} \lambda_1^x}{x!} + a_2 \frac{e^{-\lambda_2} \lambda_2^x}{x!} \quad \text{for } x = 0, 1, 2, \ldots \quad (3.9)$$

where a_1 and a_2 represent the weights associated with each of the component processes. This is a more general approach to the study of heterogeneous situations and consequently it is given the name *general double Poisson* distribution. In appendix 3.B (at the end of this chapter) a method for obtaining a_1, a_2, λ_1, and λ_2 is given. The general double Poisson distribution is discussed by Schilling (1947), Haight (1959) and Hinz and Gurland (1967). In a

Table 3.3. *The double Poisson process model: observed and*
expected frequencies based on pattern shown in fig. 3.3

Points per quadrat	Observed frequency top 25 quadrats	Observed frequency bottom 25 quadrats	Observed frequency total	Expected frequency
0	0	4	4	3.70
1	1	6	7	8.50
2	5	6	11	10.50
3	7	5	12	9.55
4	4	3	7	7.25
5	2	1	3	4.85
6	1	0	1	2.90
7	4	0	4	1.55
≥ 8	1	0	1	1.20
(N)	25	25	50	50

$X^2 = 0.994$ $df = 2$ $\chi^2_{0.20} = 3.22.$

series of articles Dacey (1965, 1966*a, b, c*) discusses various heterogeneous situations.

As in the superimposition cases, it is possible to consider probability functions other than the Poisson. Clearly, if many random variables are to be combined, models must be developed which represent these complex situations as simply as possible. For this reason we turn now to the multiplicative models which are designed to describe more varied spatial processes.

3.2.   Multiplicative processes

There are two kinds of multiplicative models explored here. The first, called *compound models,* represent the same heterogeneous situation as considered in the last section. The difference lies in the way the models are constructed. In the additive case, several random variables, each associated with a probability function representing an independent location process, were combined into one probability model. Here one probability function represents the heterogeneous nature of the location process and another represents the location process within the differing subareas. The two random variables associated with these probability functions are combined

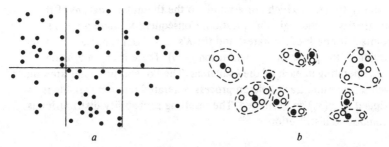

Fig. 3.4 Patterns resulting from multiplicative processes. (*a*) Compound.
(*b*) Generalized

into one by multiplication. In fig. 3.4 one can see that the type of pattern
accounted for is no different from the patterns resulting from the set of
assumptions of the heterogeneous models discussed previously. The compound models described here, however, can be thought to be more efficient
in use. Previously it was required that there be one probability function for
each subarea. Here only two probability functions are needed, no matter
how many subareas are being considered.

The second group of multiplicative models are called *generalized models*.
Each generalized model is constructed by mixing the random variable of a
probability function representing cluster size (clusters of points) with the
random variable of a probability function depicting cluster location. A typical
pattern accounted for by a generalized model is shown in fig. 3.4.

3.2.1. Compound models

As has been said, a compound model represents a location process
in an heterogeneous region, i.e. in instances where individuals have dissimilar
propensities to locate in the subareas making up the region. Some subareas
may provide more favorable environments than others so that the expected
number of individuals (parameter λ) varies from subarea to subarea. Thus, λ
is itself a random variable.

The compound model is a mixture of a random variable associated with
a probability function representative of a location process and one associated
with a probability function representing the density of points in the many
subareas of the study region. Each of the two processes is assumed to be
independent of the other; thus their mixture can arise by multiplication.
Patterns of population location, store location, and plant and animal location
can fall into this category. The spatial occurrence of thunderstorms or of
stream intersections in varied topographic conditions are further examples
of heterogeneous situations.

In discussing these models it is helpful to use the assumptions of the
Poisson process model as the location process and compound it with a

probability function which corresponds to the theoretical nature of the heterogeneity of the region in question. Consequently, there is a density of points λ for each of j subareas and the λ's vary from subarea to subarea according to the probability distribution $g(\lambda_j)$. There is also a Poisson process operating in each of the j subareas. The Poisson processes are similar in each subdivision. Each Poisson process generated random variable is multiplied by $g(\lambda_j)$ and summed. The resulting probability distribution is the compound distribution

$$P(x) = \sum_j g(\lambda_j) P_j(x; \lambda_j) \tag{3.10}$$

The choice of $g(\lambda_j)$ depends on the relative frequency of the j number of λ point density values, independent of location.

A particularly useful probability function for representing the heterogeneity is the gamma distribution. The two parameters of the gamma distribution can be set at various positive values, each pair of which describes a different distribution curve. Many of the possible gamma distribution curves are typical of the kinds of probability distributions which represent λ_j values. This flexibility has its dangers, however. It is important to be aware of the way in which the parameters describe the distribution. It becomes a meaningless exercise to use the gamma distribution unless the parameters chosen or derived from the data conform to the assumptions of the researcher.

The gamma distribution may be written as

$$g(\lambda; k, \theta) = \frac{1}{\theta^k \Gamma(k)} \lambda^{k-1} e^{-\lambda/\theta} \tag{3.11}$$

where k and θ are parameters greater than zero and λ is the random variable representing the frequency of different point densities in subareas. The symbol Γ refers to the gamma function and is defined as

$$\Gamma(k) = (k-1)! \tag{3.12}$$

When k is a positive integer there is no problem finding $\Gamma(k)$, but one must consult a table of gamma functions when k is not an integer (see Abramowitz and Stegun, 1965). Fig. 3.5 shows that as k becomes large and θ small the curve representing the gamma distribution becomes more like a normal curve. Note the tail of the distribution is elongated when k is low. An elongated tail implies that many subareas have low point densities while relatively few have high densities. A pattern created from such a process model will contain points which appear greatly clustered.

Now, returning to (3.10), substitute a gamma probability function for $g(\lambda_j)$ and a Poisson probability function for $P_j(x; \lambda_j)$. Since the gamma distribution is continuous rather than discrete it is necessary to integrate

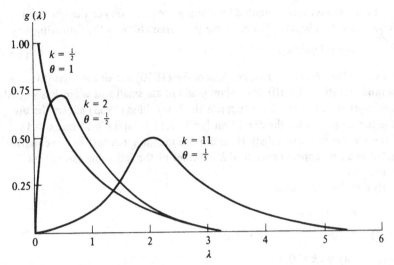

Fig. 3.5 The gamma distribution

$g(\lambda; k, \theta)$ from 0 to ∞ with respect to λ. We get

$$P(x; k, \theta) = \frac{\theta^{-k}}{x!\Gamma(k)} \cdot \frac{\Gamma(x+k)}{[(1+\theta)/\theta]^{x+k}} \quad \text{for } x = 0, 1, 2, \ldots \quad (3.13)$$

which is a negative binomial distribution with parameters k and θ. This means that a compound gamma–Poisson distribution is the same as a negative binomial distribution. The name 'negative binomial' is given since one could obtain the same distribution by expanding

$$\left[\frac{\theta}{1+\theta} + \frac{1}{1+\theta} \right]^{-k} \quad (3.14)$$

which is a binomial term with a negative exponent.

The mean of the distribution is $\mu = k\theta$ and the variance is $\sigma^2 = [\mu + (\mu^2/k)]$. Substituting μ for $k\theta$ in (3.13) and rearranging we get

$$P(x; k, \mu) = \frac{\Gamma(k+x)}{x!\Gamma(k)} \left(\frac{\mu}{k} \right)^x \left(\frac{k}{k+\mu} \right)^{k+\mu} \quad \text{for } x = 0, 1, 2, \ldots \quad (3.15)$$

The parameter k is a measure of spatial clustering. That is, low k values imply great clustering and high k values mean little clustering; thus k is consistent with its role in the gamma distribution where low k values tended to skew the distribution exemplifying great heterogeneity. The θ of the gamma distribution is equivalent to μ/k in the negative binomial distribution. High values of θ mean a tendency toward skewness in the gamma distribution; thus if k is small relative to the mean the right side of the negative binomial distribution will similarly be extended.

The parameter μ is estimated by finding the mean (m) of the observed frequencies. It is best to first estimate the parameter k in the following way:

$$\hat{k} = m^2/(v - m) \qquad (3.16)$$

where v is the observed variance. Anscombe (1950) has shown that this method is statistically efficient when k and m are small and when k is a good deal larger than m (which is often not the case). Bliss (1953) allows the use of equation (3.16) for the condition $[(k + m)(k + 2)]/m \geqslant 15$ but $k > 1$. If the k value does not satisfy these conditions then it is best to use equation (3.16) as a first approximation of k in either of the following estimation techniques.

(i) For the conditions

(1) $m < 1$, any k or

(2) $1 < m < 2, 0.4 < k < 1$ or

(3)   any $m, k < 0.4$

adjust k until both sides of the following equation balance:

$$k \log \left(1 + \frac{m}{k}\right) = \log \frac{N}{f_0} \qquad (3.17)$$

where N is the total number of observations (quadrats) and f_0 is the observed frequency of sample areas having no points.

(ii) For all other conditions use the following maximum likelihood technique. Find a k which results in a y value of zero in the following equation:

$$y = \sum_{x=0}^{n} \left(\frac{A_x}{k + x}\right) - N \ln \left(1 + \frac{m}{k}\right) \qquad (3.18)$$

There are n indexes of the random variable (for example in table 3.4, $n = 17$). A_x is the accumulated observed frequency down to $x + 1$ starting with n; N is the total number of observations. For example, in table 3.4 when $x = 13$, the A_{13} value is 8, i.e. the accumulated frequency from $x = 17$ to $x = 14$ is 8.

Further discussion of these techniques can be found in Anscombe (1950) or Johnson and Kotz (1969).

As an example of a heterogeneous Poisson process of the compound variety suppose that due to variations in the forces giving rise to the location of settlements the state of Pennsylvania can be divided into 78 equal-sized regions. The pattern of towns of over 1000 population is assumed to be located by a Poisson process in each region, however the density of towns is assumed to vary from region to region. It is therefore assumed that a negative binomial process model is an appropriate model for the location of settlements of 1000 or more people for the entire state. The assumptions

Table 3.4. *Compound negative binomial process model: observed and expected frequencies of the location of towns of 1000 population or more in Pennsylvania in 1970 by quadrats*

Points per quadrat	Observed frequency	Expected frequency
0	8	8.424
1	11	8.892
2	5	8.424
3	9	7.566
4	12	6.708
5	4	5.772
6	3	4.992 ⎱
7	2	4.290 ⎰
8	2	3.666 ⎱
9	4	3.120 ⎰
10	4	2.652 ⎱
11	1	2.184 ⎰
12	1	1.872 ⎰
13	4	1.560 ⎱
14	1	1.326 ⎱
15	1	1.092 ⎰
16	2	0.936 ⎰
≥17	4	4.524 ⎰
(N)	78	78

$X^2 = 9.74$ $df = 7$ $\chi^2_{0.20} = 9.80$

about the nature of the gamma distribution will not be discussed, but it should be made clear that the theory being tested concerns the nature of the gamma distribution and the assumptions of a Poisson location process.

On fig. 3.6 the number of settlements is shown for each region. For brevity no tests on the Poisson process assumptions are made. Table 3.4 gives the results of the negative binomial test. In the example m is 5.962; the k value, 1.29, was estimated by (3.18). With m and k, each of the expected frequencies can be calculated. When $x = 0$ the following equation is used:

$$P(0; m, k) = 1/[(k + m)/k]^k \qquad (3.19)$$

and subsequent x values are determined by

$$P(x; m, k) = [(k + x - 1)/x] \, [m/(k + m)] \, [P(x - 1)] \qquad (3.20)$$

where $P(x - 1)$ is the expected proportion calculated for the value $(x - 1)$. That is, to find the expected probability $P(x = 4)$, the value for $x = 3$ is needed.

In the experiment just described the table value was greater than the

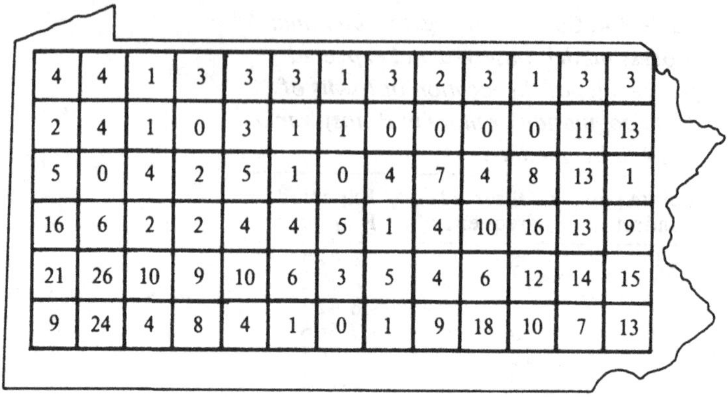

4	4	1	3	3	3	1	3	2	3	1	3	3
2	4	1	0	3	1	1	0	0	0	0	11	13
5	0	4	2	5	1	0	4	7	4	8	13	1
16	6	2	2	4	4	5	1	4	10	16	13	9
21	26	10	9	10	6	3	5	4	6	12	14	15
9	24	4	8	4	1	0	1	9	18	10	7	13

Fig. 3.6 Towns of population greater than 1000 in Pennsylvania in 1970 by quadrats

observed X^2 value of 9.74. For df = 7 the value at the 0.20 level of significance is 9.80, thus we can conclude that the negative binomial might be an appropriate model for the spatial pattern of settlements in Pennsylvania.

If the pattern were studied at different time periods then it would be possible to make statements about the nature of pattern changes by evaluating changes in the value of k. Increased grouping will lower the value of k. Some researchers use $1/k$ as an index of grouping so that increases in grouping will correspond to increases in the index. The value of k will remain the same over time even if the number of points decreases as long as the deaths are randomly selected. Similarly, if births are located according to a Poisson process the k value will not change. This property of the k parameter is valuable in assessing how change takes place. The parameter must not be used for this purpose, however, if the distribution is not a negative binomial.

Let us explicitly list the assumptions needed for the heterogeneous negative binomial model discussed above.
(1)   Within each subarea each possible location is equally likely to receive a point.
(2)   The location of any point in a subarea is independent of the location of any other point.
(3)   The frequency of the various densities of points in subareas is distributed as the gamma distribution with parameters θ and k. (The parameters θ and k may or may not be specified *a priori*).
Note that nothing has been said about the point densities of subareas relative to their location. If it were assumed that the number of points in one subarea is independent of the number of points in any other subarea we would not expect a correlation between densities of contiguous subareas (no spatial autocorrelation). Not having made the independence

assumption allows for spatial autocorrelation. This is a useful omission since autocorrelation often occurs in empirical work.

Dacey (1968) discusses this problem in detail in his study of house location patterns in Puerto Rico. He shows, for example, that a cluster of points may be bisected by the boundary between subareas which results in a violation of the first assumption. In such cases, the researcher should attempt to select a subarea (quadrat) size where this is minimized. This might best be done by delimiting clusters of points beforehand, and then choosing a quadrat size. In this context Getis (1969) and Boots and High (1974) have suggested that the grouping criteria conform to the nature of the subject matter represented by the points.

A confusing aspect of the use of the negative binomial distribution is that it can be developed from rather different circumstances than those just discussed. While multiple origins are true of most probability distributions, the geographic literature has suffered particularly in the case of the negative binomial distribution. There are many instances when it could have been used, but some researchers have either felt that they were not in a position to postulate the necessary assumptions or they confused the heterogeneous model with other models. Harvey (1966, 1968) has explored these problems.

Perhaps the work of Artle (1965), and Rogers (1965), and Lee (1974) would best be interpreted as models of heterogeneous populations. In all of these studies the pattern of stores within cities were assumed to result from a negative binomial distribution. Others who have demonstrated its use are McConnell (1966) and Susling (1971b). Tables of the negative binomial are provided by Williamson and Bretherton (1963). Further bibliographic notes and comments on the discrimination between different negative binomial models are provided in the next section.

A Poisson process generated random variable may be compounded with any one of a number of random variables representing probability functions other than the gamma. In some circumstances it might be well to use a Poisson as the probability function representing the heterogeneity. This would be a Poisson–Poisson compound distribution. We will discuss such a distribution in the next section on generalized models under the heading Neyman Type A distribution. Johnson and Kotz (1969) and Gurland (1958) describe such models as Poisson–negative binomial, Poisson–logarithmic series and Poisson–hypergeometric, and also several distributions having a non-Poisson as the location probability function and the Poisson as the probability function representing heterogeneity. Among these are the binomial–Poisson and the negative binomial–Poisson (known as the Poisson–Pascal distribution). The set of assumptions is different for each model, thus each can represent a particular pattern-generating process. Hopefully with more research on the nature of spatial processes the confusion will subside about which model to use.

The negative binomial distribution is discussed in great detail by Anscombe (1950) and compared with other similar distributions. The work of Fisher (1941) and Skellam (1948) laid the groundwork for much of what followed. Johnson and Kotz (1969) review a good deal of the literature on the subject. Of special interest is Boswell and Patil's (1970) survey of the myriad ways one might derive the negative binomial. Rogers (1969a, 1974), Ord and Patil (1972), Gurland (1958) and Anscombe (1950) clarify the difference between the compound and generalized negative binomial models. These will be discussed in the section on generalized models which follows.

3.2.2. Generalized models

Generalized models deal with patterns of groups of points rather than with individual points. A generalized distribution is a combination of a random variable of a probability function describing group size and a random variable of a probability function describing group location. Two examples are:

(1)   A model for the pattern of houses in a rural area. The model would be a mixture of two random variables, one associated with a probability function accounting for the location of villages and the other with a probability function describing the number of houses in the villages.

(2)   A model for the spatial pattern of adopters of an innovation or idea. The model would be a mixture of a probability function accounting for the location of those who diffuse the idea and a probability function describing the number of adopters related to each of the individuals who diffuse the idea (contagion model).

Clearly, models that can express the spatial manifestation of complex processes can hardly show the causal connections between process and spatial form. The process models selected, however, can embody much in the way of theory regarding the organization of individuals in space. For example, theories of village formation, innovation diffusion, plant and animal life cycle characteristics, human interaction, and so on may be tested in part by using appropriate generalized models. These are models of dependency and therefore are based on the existence of conditional distributions.

The generalized models shown here are designed to account for the number of points per sampling area. The mixture is represented by the conditional distribution $P(x|y)$ (which reads 'the probability of x individuals given y clusters') and a probability function representing the number of clusters $P(y)$ (Ord and Patil, 1972). Thus,

$$P(x) = \sum_y P(x|y) P(y) \qquad (3.21)$$

This general expression means that the probability of x points in a sample area is obtained by multiplying a proportion representing the expected number of y clusters in the sample area by the expected proportions for each possible number of points per cluster and summing.

Neyman Type A process model. Let us substitute into (3.21) a Poisson process generated random variable to represent clusters per sample area and a Poisson process generated random variable to represent points per cluster. We get

$$P(x; \lambda_1, \lambda_2) = e^{-\lambda_1}\left[\lambda_1 e^{-\lambda_2} \frac{\lambda_2^x}{x!} + \frac{(\lambda_1 e^{-\lambda_2})^2}{2!}\frac{(2\lambda_2)^x}{x!} + \frac{(\lambda_1 e^{-\lambda_2})^3}{3!}\frac{(3\lambda_2)^x}{x!} + \dots\right] \quad \text{for } x = 0, 1, 2, \dots \quad (3.22)$$

If we let j equal the number of points per cluster, and let it vary over the range 0 to ∞ we obtain

$$P(x; \lambda_1, \lambda_2) = e^{-\lambda_1}\frac{\lambda_2^x}{x!}\sum_{j=0}^{\infty}\frac{(\lambda_1 e^{-\lambda_2})^j}{j!} \cdot j^x \quad \text{for } x = 0, 1, 2, \dots \quad (3.23)$$

This then becomes a Poisson–Poisson generalized distribution or, as it is more commonly called, the Neyman Type A distribution after its originator Jerzy Neyman (1939). Neyman's problem was to test a theory related to the number of larvae of insects on pine shoots. Implicit assumptions are that each cluster of larvae is distinct and that no sampling area boundary cuts across a cluster. These assumptions suffice when clearly disjoint sampling units are used, but problems do develop when use is made of quadrats whose boundaries may cut across a cluster. Before discussing this problem let us first complete our discussion of the nature of the Neyman Type A.

The mean is $\lambda_1\lambda_2$. The variance is $\lambda_1\lambda_2 (1 + \lambda_2)$, which obviously exceeds the mean, thus indicating that the points in a pattern that is the realization of a Neyman Type A process are more clustered than those in a pattern resulting from a Poisson process. The parameters may be estimated from the sample data by:

and
$$\left.\begin{array}{l}\hat{\lambda}_1 = m_1 = m/m_2 \\[2mm] \hat{\lambda}_2 = m_2 = (v - m)/m\end{array}\right\} \quad (3.24)$$

where m is the mean and v is the variance.

Since j must be summed to infinity a special method for finding the various values of $P(x; \lambda_1, \lambda_2)$ needs to be followed. One procedure requires the use of a table supplied by Douglas (1955), but barring its availability the following technique is useful. First determine m_1 and m_2 and find

$$P(0) = \exp[-m_1 (1 - e^{-m_2})] \tag{3.25}$$

which is the expected proportion of 0 occurrences.

For $x > 0$ the following expression is used:

$$P(x) = \frac{m_1 m_2 \, e^{-m_2}}{x} \sum_{j=0}^{x-1} \frac{m_2^j}{j!} P(x-j-i) \quad \text{for } x = 1, 2, \ldots \tag{3.26}$$

Thus,

$$P(1) = m_1 m_2 \, e^{-m_2} P(0)$$

$$P(2) = \frac{m_1 m_2 \, e^{-m_2}}{2} [P(1) + m_2 P(0)]$$

$$P(3) = \frac{m_1 m_2 \, e^{-m_2}}{3} [P(2) + m_2 P(1) + \frac{m_2^2}{2!} P(0)]$$

$$P(4) = \frac{m_1 m_2 \, e^{-m_2}}{4} [P(3) + m_2 P(2) + \frac{m_2^2}{2!} P(1) + \frac{m_2^3}{3!} P(0)]$$

and so on.

In geography the Neyman Type A process has been used to study the clustering of adopters of an innovation (Harvey, 1966), the location of vacation houses in an area of Sweden (Aldskogius, 1969), and the location of retail stores in several cities (Rogers, 1965, 1969a). In all cases the researchers were unwilling to accept the Neyman Type A as the appropriate model. The difficulty lies in the nature of the assumptions. Let us review these.

Poisson location process:

(1)   Within a region each possible location is equally likely to receive a cluster.

(2)   The placement of any cluster is independent of the placement of any other cluster.

(3)   The parameter λ_1 corresponds to the *a priori* belief in the density of clusters (optional assumption).

Poisson cluster process:

(4)   The parameter λ_2 corresponds to the *a priori* belief in the mean size of clusters (optional assumption)

(5)   In conformity with a Poisson model the variance about the mean size of cluster is equal to the mean size of cluster.

(6)   The assignment of a cluster size (points) to one cluster location is independent of the assignment of any other cluster size or cluster location (optional assumption).
(7)   The points in the cluster are propagated by a 'progenitor' who is located at the site of each cluster (pseudo-contagion assumption).

If the Neyman Type A process is chosen as the model then one must be satisfied with the Poisson assumptions which are very simple indeed. It is helpful, but not necessary, for the researcher to specify expected λ values before the observed frequencies are compared to the theoretical Neyman Type A values. The expected values depend on *a priori* knowledge of the nature of the location and clustering processes. As has been mentioned earlier, when it is possible to specify parameters it is not necessary to estimate them from observed data.

If use is made of assumption 6 but not 7 we would be describing a heterogeneous model of the compound type. By using 7 we create a contagion process, but (3.23) does not make this explicit. Thus, 7 is considered a pseudo-contagion assumption. By dropping 6 we imply that there may be an association between the sizes of neighboring clusters. If the association is pronounced there is the possibility that the Neyman Type A model would be rejected on the grounds of overly large numbers of points per quadrat. If a lattice of quadrats is used for testing purposes there is the problem of contiguous quadrats having large numbers of points (or small numbers of points). This should not be troublesome if care is taken not to have quadrat boundaries bisect clusters. More will be said about the subject of spatial autocorrelation later in this chapter.

The 'contagious' interpretation given this and other generalized distributions is troublesome. Here 'contagious' implies a process with at least two stages. The first stage is a location process (Poisson process in the case of the Neyman Type A model) simulating the site of the progenitors and the second stage is an 'infectious' stage where each progenitor 'creates' a cluster of related individuals. It must be realized that by using a Poisson process generated random variable for cluster size, some clusters would be assigned the value 0 which means no members. For example, the distribution of people in houses may be assumed to follow a Neyman Type A model; on occasion nobody would be present in a house and this would be interpreted as a cluster of zero size. For most problems, however, each Poisson location is assumed to have a cluster size of at least one (which can be interpreted as at least containing the 'originator'). The assignment then would follow a truncated Poisson process.

Differences between the Poisson and truncated Poisson distributions for a given point density are small enough so that the Neyman Type A model is representative of both. The estimate of the parameter λ_1 is different, however.

The parameter λ_2 remains the same. For the truncated case:

$$\hat{\lambda}_1 = m_1 = \frac{m}{m_2}\,(1 - e^{-m_2}) \tag{3.27}$$

This follows from the fact that the expected mean size of a cluster using the truncated Poisson distribution is $m_2/(1 - e^{-m_2})$ which is larger than m_2 of the non-truncated case.

Table 3.5. *Neyman Type A process model*
(Poisson–truncated Poisson process): observed
and expected frequencies based on simulated
data shown in fig. 3.7.

Points per quadrat	Observed frequency	Expected frequency
0	2	5.300
1	4	1.525 ⎫
2	2	2.325 ⎬
3	2	2.575 ⎭
4	2	2.425 ⎫
5	5	2.125 ⎬
6	3	1.825 ⎭
7	0	1.575 ⎫
8	3	1.400 ⎬
≥9 (15, 18)	2	3.925 ⎭
(N)	25	25

$X^2 = 5.025 \qquad df = 1 \qquad \chi^2_{0.20} = 1.642$

Table 3.5 gives the results of the experiment shown in fig. 3.7 using the random variable of table 2.6 as the generalizing distribution. Since it was decided to use all of the assumptions listed above, the assignment of cluster size was accomplished in the following way. Each of fifty points were placed in the study area in a Poisson process manner and numbered from one to fifty. A three-digit random number was associated with each of the points. If the random number fell between 000 and 244 (from table 2.6 it is evident that an expected proportion of 0.245 is representative of outcome 1) then the cluster size would be one. Random numbers 245–536 represented outcome 2, and so on. As a result 50 groups ranging in size from 1 to 6 were developed. The *total* number of points by quadrats in the sample area is the realization of a Poisson process.

The parameters were estimated as $m_1 = 1.655$ and $m_2 = 2.762$. Note that the Neyman Type A distribution can have more than one mode. In our example there is a peak at $x = 0$ and $x = 3$. The mean observed cluster size is 2.44, which is less than the expected value of $m_2/(1 - e^{m_2}) = 2.95$. The fact that one quadrat has 15 individuals and another 18 creates a high variance

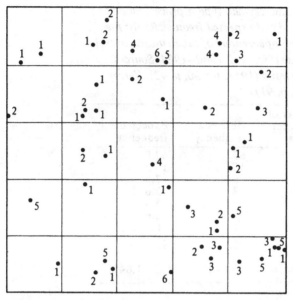

Fig. 3.7 Pattern resulting from Neyman Type A process. Each point represents a cluster of size indicated

value (18.36) as compared to the mean (4.88) and this results in a high expected cluster size.

Perhaps a more helpful example is one provided by Harvey (1966) who used data due to Hägerstrand (1953). Table 3.6 summarizes the results of a test on the location of acceptors of tuberculosis control in cattle during the period 1900–1924 in a rural area of Sweden. For the test, one assumes that the location of the acceptors is due to a Poisson process and then clusters of new acceptors develop around the original acceptors according to a Poisson process. The entire process is assumed to be a Neyman Type A process. The truncated Poisson process was not used, indicating that some acceptors did not disseminate information about tuberculosis control, or at least were not able to generate a group of acceptors, and in fact by the end of the period were not acceptors themselves. The results show a similarity between the observed and expected frequencies indicating that the assumptions may be correct.

Thomas process model. Another way to insure that the originator is included in the pattern is to hypothesize a Thomas process. The mean number of clusters per quadrat is given by the parameter λ_1 and the objects per cluster are described by $1 + \lambda_2$. Consequently, λ_2 is one less than the mean cluster size, but $1 + \lambda_2$ explicitly takes into account situations in which a 'parent' and 'offspring' are included in the count for each cluster. Thomas's process

Table 3.6. *Neyman type A process model:*
observed and expected frequencies for the
location of accepters of tuberculosis control
in cattle (Sweden), 1900–1924. Sources:
Hägerstrand (1953, fig. 30, p. 92), Harvey
(1966, p. 91)

Number of points per quadrat	Observed frequency	Expected frequency
0	63	62.8
1	15	13.2
2	6	7.9
3	3	3.6
4	1	1.4
5	1	0.6
≥ 6	1	0.5
(*N*)	90	90

$X^2 = 0.705$ $df = 1$ $\chi^2_{0.20} = 1.642$

model accounts for the number of objects per quadrat. The parameters are
found by

$$\lambda_1 = \frac{\mu}{1 + \lambda_2} \tag{3.28}$$

where μ is the mean number of points per quadrat (which is estimated by m,
the sample mean), and $1 + \lambda_2$ is equal to the assumed number of points per
cluster. The distribution is written

$$P(x; \lambda_1, 1 + \lambda_2) = \sum_{j=1}^{x} \frac{\lambda_1^j e^{-\lambda_1}}{j!} \frac{(j\lambda_2)^{x-j} e^{-j\lambda_2}}{(x-j)!} \quad \text{for } x = 1, 2, 3, \ldots \tag{3.29}$$

where we let j range from 0 to ∞. It is interesting to note that

$$P(0; \lambda_1, 1 + \lambda_2) = e^{-\lambda_1} \tag{3.30}$$

For a discussion of the Thomas distribution see Thomas (1949) and Pielou
(1957).

In practice each of the individuals in a cluster is represented by a point.
When this is the case, it must be remembered that the Neyman Type A or the
Thomas process is not concerned with the *arrangement* but only with the
number of individuals within the cluster. Clusters must be thought of as point
locations, consequently it is important that sampling area boundaries do not
cut through clusters. Also, if clusters overlap, it is important that cluster mem-
bership be defined carefully.

Pielou (1957) points out that quadrat size can greatly affect results. She makes a most important point by suggesting that more than one quadrat size be used. The benchmark for quadrat size should be that the area is chosen in regard to the mean cluster area.

For both the Neyman Type A and the Thomas processes we expect to find that the point location of groups follow a Poisson process. Therefore it may be useful to identify the groups and test for Poissonness by measuring distances to centers (centroids) of nearest neighboring groups. (For a discussion on finding the center of a pattern of points see Neft, 1966.) This implies that quadrat sampling may not be necessary at all in order to verify the model. Getis (1969) suggests that if a lattice is used in quadrat sampling the orientation of the sampling areas should be changed for each experiment but caution should be used because of the possible correlation between experimental results. Rogers and Gomar (1969, p. 382) propose that 'the optimal quadrat size is that size which maximizes the power of the chi-square test for a given level of significance'. By taking the variance − mean ratio (degree of clustering) the null hypothesis of a Neyman Type A process is evaluated in terms of the negative binomial distribution (the alternative hypothesis) for different quadrat sizes.

A literature on the identification of groups has arisen recently. Hamilton (1971a), Dacey (1969c), Roach (1968), Ling (1972, 1973a) and Boots and High (1974) have suggested various methods of classification. Boots and High develop a critical distance method in which points are combined into groups if nearest neighbor distances are less than the mean distance to nearest neighbor for the entire region. Using a slightly different method of cluster identification Roach (1968) develops a way in which probabilities for various cluster sizes may be found. Lee (1972) following Roach considers the size of clusters of settlements. Dacey (1973) studies clustering in a related, but more comprehensive model approach. He uses a single spatial focus around which cluster centers are located, and within a cluster, points are placed following the assumptions of a circular symmetric process. This approach is discussed in a series of articles where the patterns of retail and service establishments in an urban area are the data explored (Dacey, 1969a, b).

In a related development, Matérn (1960) introduces a doubly stochastic Poisson process which is a two-stage area−point process resulting in patterns in which individual points cluster. First, circles of fixed size are dropped over the plane according to a Poisson process. Those areas of the plane covered by i circles have a probability p_i of receiving a point. For each area of type i points are placed according to a Poisson process having density λ_i. In this regard the center−satellite process of Neyman and Scott (1958) and Bartlett's (1964) paper on the spectral analysis of two-dimensional point processes should be mentioned as additional examples of clustering process models. An area model developed by Getis and Jackson (1971) which has much in common with the Matérn approach is discussed in chapter 7.

Generalized models, including the contagious distribution interpretations of them, are discussed in detail by Johnson and Kotz (1969), Neyman (1939), Feller (1943), Skellam (1958), Gurland (1958) and many others. In the text by Pielou (1969) much space is devoted to their explanation for biologists. A clear discussion of the differences between the various generalized models is given by Ord (1972). Technical literature of great help on sampling, parameter estimation, and testing has been prepared by Shenton (1949), Martin and Katti (1965), Katti and Gurland (1962), Douglas (1955), Rogers and Raquillet (1972) and Rogers and Gomar (1969).

Negative binomial process model. Instead of assuming that group sizes are the realization of a Poisson process, let us now assume that a process creating a highly skewed frequency of group sizes is responsible for the clustering process. In many instances this is a realistic assumption to make. Theories of plant, animal, and human organization usually call for a highly skewed distribution of frequencies with few cases of many members per group and many cases of few members per group. The logarithmic series distribution (LSD) model is useful in this regard. It is written

$$P(x;\alpha) = \frac{-1}{\ln(1-\alpha)} \frac{\alpha^x}{x} \quad \text{for } x = 1, 2, 3, \ldots \quad (3.31)$$

The mean and variance are

$$\mu = \alpha/\{-[\ln(1-\alpha)](1-\alpha)\} \quad (3.32)$$

$$\sigma^2 = \mu\{[1/(1-\alpha)] - \mu\} \quad (3.33)$$

There is no zero category thus making it unnecessary to truncate the distribution. An estimate of α may be obtained from the observed data by finding the mean (m) and using the table provided by Williamson and Bretherton (1964) or by trial and error using (3.32).

As an illustration of the LSD model, sizes of clusters of houses in Piscataway Township, New Jersey, in 1850 were thought to be the realization of a logarithmic process (Getis, 1969). Table 3.7 gives the results of a test on that hypothesis. The observed mean is 2.922 and $\hat{\alpha}$ is found to be 0.8448. The largest numerical difference between observed and expected frequencies is in the category of one indicating that more single houses were in evidence than expected. The observed X^2 value of 6.101 is within the tabled value of 7.289 at the 0.20 level of significance for $5 df$ allowing us not to reject the LSD model.

In order to develop a generalized model according to (3.21) the two

Table 3.7. *Logarithmic series process model:
observed and expected frequencies of clusters
of houses in Piscataway Township, New
Jersey in 1850*

Houses per cluster	Observed frequency	Expected frequency
1	69	58.566
2	15	24.768
3	12	13.932
4	9	8.772
5	8	5.934 ⎫
6	3	4.257 ⎭
7	1	3.096 ⎫
8	4	2.193 ⎭
9	2	1.677 ⎫
10	1	1.290
11	1	1.032
≥12 ⎡14⎤⎢15⎥⎢17⎥⎣24⎦	4	3.483 ⎭
(N)	129	129

$X^2 = 6.101$ $df = 5$ $\chi^2_{0.20} = 7.289$

distributions, logarithmic series and Poisson, are mixed. The result is

$$P(x; k, m) = \frac{\Gamma(k + x)}{x! \Gamma(k)} \left(\frac{m}{k}\right)^x \left(\frac{k}{k + m}\right)^{k+x} \quad \text{for } x = 0, 1, 2, \ldots \quad (3.34)$$

which is the negative binomial distribution of (3.13) and (3.15). The α of the LSD model is equal to $m/(k + m)$ which indicates how the clustering parameter k affects the nature of the negative binomial model. The value of k is estimated from any of formulas (3.16, 17, 18). When k is low relative to m, great clustering is implied. The curve representing the LSD under those circumstances would have an elongated tail reflecting the existence of many clusters. When k is high relative to m, the α of the LSD is low indicating that a large proportion of the expected frequencies is assigned to the one category. It should be made clear that the θ of the gamma distribution (3.11) is equal to m/k of the negative binomial distribution and is not to be confused with the α of the LSD. In treatments of mathematical statistics, it is shown that the logarithmic series distribution becomes the negative binomial distribution when k is allowed to go to zero. If k is made very large a Poisson distribution results (Quenouille, 1949; Johnson and Kotz, 1969). Thus it can be seen that it is the expected degree of clustering that is the link between the models (Rogers, 1969a).

What we have shown is that by compounding the gamma probability function with the Poisson probability function we produce a heterogeneous model which looks exactly like the generalized distribution arrived at by mixing the logarithmic with the Poisson probability function. Although the two equations are the same, we emphasize that the heterogeneous model is different from the generalized model. Whereas the k parameter in both models describes the nature of clustering, in the generalized model the implication is that the k represents the tendency for the clustering of individuals into groups. The compound model assumes a heterogeneous surface in which each subarea displays a pattern which is the realization of a Poisson process. The generalized model assumes a homogeneous surface, where each subarea is equally likely to receive points but each point has associated with it other points. The number of associated points is the result of a logarithmic process.

As a result of these differences in the model assumptions, the manner in which tests are carried out is different. In the case of the heterogeneous model quadrats of appropriate size are useful sampling devices. Since there is no specified way that the realization of a gamma process is assumed to vary spatially, the quadrat approach (either a lattice of quadrats or randomly placed quadrats) is adequate. As a check, one should be convinced that each quadrat contains a set of points which could have resulted from a Poisson process.

In the case of the generalized model, it may be appropriate to disregard a test on the negative binomial hypothesis and instead substitute two tests — tests on the Poisson and logarithmic series models. If the researcher's expectations are satisfied by tests on each of these (neither is rejected), he can conclude that the negative binomial model may very well be appropriate. He may then use a quadrat approach in order to ascertain the parameter estimates and the moments of the negative binomial distribution.

The generalized negative binomial model may represent a dynamic growth (contagious) situation. It may be assumed that the cluster centers are located at uniform time intervals. As each center is placed, a cluster begins to develop. If the birth rate of cluster members is constant, the resulting pattern at any time period is the realization of a negative binomial process. One may work the other way as well, that is, assume an already existing pattern of clusters which suffers from a constant death rate. Here again the total pattern is the realization of a negative binomial process and the cluster sizes will be distributed as the LSD.

Much has been written about the negative binomial distribution, some of which has been referred to in the previous section on compound models. In its generalized context biometricians such as Pielou (1969), Bliss (1953), Bliss and Owen (1958), and Anscombe (1950) have used it and made contributions to its theory. In geography Dacey (1968) studied the pattern of houses in Puerto Rico and found it difficult to assume independence among sampling areas. Aldskogius (1969) and Getis (1969) also looked at settlement patterns. Harvey (1966) considered the contagion implied by the Hägerstrand data while Getis (1967) attempted to distinguish between the two negative binomial models mentioned by studying aspects of the k parameter. Cliff and Ord (1973) state that while the k parameter will not change over time in a heterogeneous process model, this would not be the case in a contagious situation.

Dacey (1968) and Cliff and Ord (1973) point out how the k parameter should or should not change when adjacent quadrats in a lattice of quadrats are combined to form larger quadrats (this is the problem of spatial auto-correlation). Cliff and Ord suggest that by taking an observed point pattern and calculating estimates for $p = 1/(1 + \theta)$ and k for different sizes of quadrats ($\theta = m/k$) one can detect if one or the other model is appropriate. If a number s of the original quadrats are combined to form larger quadrats then for the generalized model the k of the original estimate should equal sk, but k should remain the same in the compound model. The effect on p is different, however. In the generalized model p should not change when quadrats are combined, but in the compound case p should equal $1/(1 + s\theta)$. What in effect is being studied is the autocorrelation which may exist between adjacent quadrats due to overlapping clusters or quadrat boundaries cutting through clusters, or other phenomena which force us to question the assumption concerning the independence of the placement of clusters. The study of autocorrelation is highly relevant to map pattern analysis and can be pursued in detail in Cliff and Ord (1973).

In statistics Boswell and Patil (1970) have shown that in addition to the negative binomial models presented here, there are at least a dozen other negative binomial models. If one can logically deduce or prove that the particular set of negative binomial model assumptions used lead to the observed pattern then there is no need to be concerned about other negative binomial models.

Polya–Aeppli process model. Another generalized model of note is the Polya–Aeppli. Here it is assumed that if progenitors are patterned as the result of a Poisson process and we observe the total number of points, after reproduction, we shall expect the number of progeny per quadrat to follow the realization of a Polya–Aeppli process. The random variable representing the expected number of points per cluster is associated with a geometric probability function. The probability distribution is

$$P(x;\delta) = (1 - \delta)\delta^{x-1} \quad \text{for } x = 1, 2, 3, \ldots \tag{3.35}$$

where δ is the parameter representing cluster size. When mixed with a

Poisson probability function the resulting distribution is

$$P(x; \lambda, \delta) = e^{-\lambda} \delta^x \sum_{j=1}^{x} \binom{x-1}{j-1} \frac{1}{j!} \left(\frac{\lambda(1-\delta)}{\delta} \right)^j \quad \text{for } x = 0, 1, 2, \ldots$$

(3.36)

The parameters are estimated by

$$\hat{\delta} = \frac{v-m}{m+v}$$

(3.37)

and $\hat{\lambda} = 2[m^2/(v-m)]$

(3.38)

The model was developed to account for certain processes of vegetative reproduction, especially freely reproducing matter. It would seem that the use of a geometric probability function would be particularly appropriate in circumstances of cluster development when no competition is present (Kendall, 1949). Some patterns of early settlement or some contagious situations may be modeled by the Polya–Aeppli process. In appendix (3.A) to this chapter the assumptions of the model are clarified by simulating a Polya–Aeppli process.

In the social sciences Harvey (1966) and Rogers (1965) have discussed the use of the Polya–Aeppli model in an experimental way. McConnell (1966) and Susling (1971*b*) have also demonstrated its use. A thorough description of the statistical properties of the distribution is given in Johnson and Kotz (1969). In a useful discussion of several generalized models Anscombe (1950) points out that the Polya–Aeppli distribution has either one or two modes. He notes that if the progenitors are released at uniform time intervals, we get the negative binomial distribution. He goes on to say that '. . . the study of counts made all at one time is not likely to give reliable information on laws of population growth. For this purpose, repeated observations on the same population are needed, if possible with identification of individuals' (p. 366). In the case of the Polya–Aeppli model it would seem useful to study the size frequency of clusters as well as the spatial pattern of clusters from time period to time period.

Appendices to Chapter 3

3.A.   **An example of the Polya–Aeppli process model**
Suppose points (progenitors) are placed following the Poisson model assumptions. Fig. 3.8 shows the location of fifty points in a region divided into 25 quadrats. Let us consider that the expected cluster size d at each of these locations is 3. A geometric probability function can be developed using (3.35). First, we get an δ value from the relationship

Table 3.8. *Polya−Aeppli process model:*
observed and expected frequencies based
on pattern shown in fig. 3.8

Individuals per quadrat	Observed frequency	Expected frequency	
0	2	2.200	⎫
1	2	2.200	⎬
2	4	2.400	⎭
3	4	2.425	⎫
4	0	2.325	⎬
5	1	2.175	⎭
6	3	1.950	⎫
7	0	1.750	⎬
8	2	1.425	⎭
9	1	1.175	⎫
10	0	1.025	⎬
11	3	0.850	
12	1	0.800	⎬
$\geqslant 13 \begin{bmatrix}16\\17\end{bmatrix}$	2	2.300	⎭
(N)	25	25	

$$X^2 = 0.867 \qquad df = 1 \qquad \chi^2_{0.20} = 1.642$$

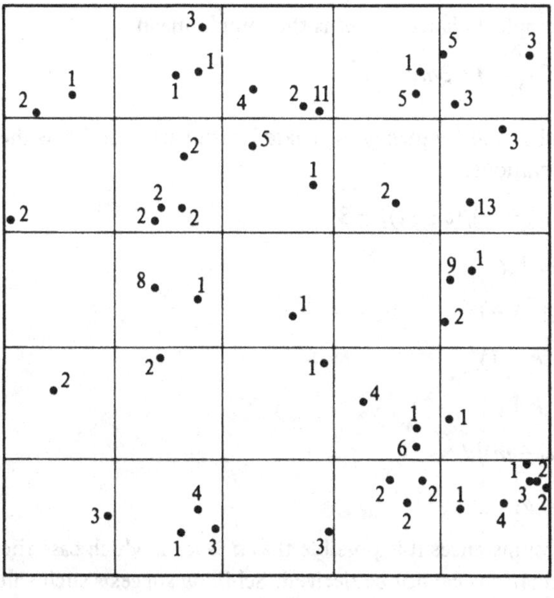

Fig. 3.8 Pattern resulting from Polya−Aeppli process. Each point represents
a cluster of size indicated

$d = 3 = 1/(1 - \delta)$. The resulting value of $\delta = 0.667$ is then substituted into (3.36). From this we obtain the proportion of clusters which are size 1, 2, 3 and so on. Random numbers are used now for the assignment of a cluster size to each point. Random numbers 001 to 333 correspond to a cluster size of 1 being assigned, 334–555 represent size 2 and so on. Now the total number of points are summed for each quadrat. These observed frequencies should be similar to those of a Polya–Aeppli model. See table 3.8 for the results. It should be noted that $\hat{\delta}$ and $\hat{\lambda}$ were estimated by (3.37) and (3.38). These values are 0.589 and 2.436, respectively. $\hat{\delta}$ is lower than δ since no single quadrat contained more than 17 points. For a probability function such as this it is expected that one or more quadrats contain a great number of points. The observed X^2 value is 0.867, well below the tabled value of 1.642 for 1 df at the 0.20 level of significance, indicating that the observed and theoretical frequencies are, as expected, very similar.

3.B.   **Finding the parameters of the general double Poisson distribution (Schilling, 1947)**

$$P(x; \lambda_1, \lambda_2, a_1, a_2) = a_1 \frac{e^{-\lambda_1} \lambda_1^x}{x!} + a_2 \frac{e^{-\lambda_2} \lambda_2^x}{x!} \quad \text{for } x = 0, 1, 2, \ldots$$

(1) $h = 4(v - m)$

where v is the sample variance and m is the sample mean.

(2) $\mu_3 = \left(\dfrac{\Sigma x^3 f_x}{N} \right) - 3mv - m^3$

where f_x is the observed frequency associated with each x and N is the total number of observations.

(3) $g = [(\mu_3 - m)/(v - m)] - 3$

(4) $k = 2m + g$

(5) $d = \sqrt{(g^2 + h)}$

(6) $\lambda_1 = \frac{1}{2}(k - d)$

(7) $\lambda_2 = \frac{1}{2}(k + d)$

(8) $a_1 = (\lambda_2 - m)/d$

(9) $a_2 = 1 - a_1$

N.B. In some circumstances it is possible that $d > k$, in which case the double Poisson expectations cannot be derived. Schilling suggests such values are indicative of the inappropriateness of a double Poisson model.

4

Truly contagious models, disturbed lattices and information theory

This chapter brings together further aspects of the theory pertaining to point pattern analysis. We extend our discussion of pattern models to include some variations on the basic themes developed in chapters 2 and 3. As an extension of the work on the pseudo-contagious models in chapter 3 we briefly present three models which may be classified as 'truly contagious'. In addition there is a discussion of two approaches to pattern study which are rooted in the work presented in chapter 2. These are disturbed lattice models and information theory.

4.1. 'Truly contagious' models

The generalized models discussed in the previous chapter have two steps, that is, they are designed to account for a point pattern after a location process and a clustering process have taken place. They can be viewed as pseudo-contagious models because they do not explicitly take into account the many stages leading to the observed spatial pattern. Models that are 'truly contagious' would in some way take the stages into account. There are at least three kinds of truly contagious models: Markov-chain models, simulations, and urn models.

4.1.1. Markov-chain models

Markov processes are used to study a variety of phenomena which concern the state of a system after a number of changes have taken place. The models have been helpful in the social sciences for the study of population change, social systems change, the spread of rumors, etc. (see Bartholomew, 1967). The Markov process that we use here assumes dependence between the determinants of point patterns in one time period and in the immediately preceding time period. The pattern as it appears after n time periods is termed the *state of the system*. A given state of the system is made up of a series of outcomes, one for each subarea of a region. Each state is linked (chained) to a preceding state by a set of probability values termed *transition probabilities*. Once the subareas are defined, the transition probabilities are applied to an initial system state and from this the behavior of the system can be computed over time. The transition probability matrix is best developed from subject-matter theory, such as friction of distance theory, migration theory, and spatial interaction theory.

Suppose there are three subareas a_1, a_2, and a_3. We can write the transition probability matrix as

$$
t \left\{
\begin{array}{c}
a_1 \\
a_2 \\
a_3
\end{array}
\right.
\overbrace{
\begin{array}{ccc}
a_1 & a_2 & a_3 \\
\end{array}
}^{t+1}
\begin{bmatrix}
p_{11} & p_{12} & p_{13} \\
p_{21} & p_{22} & p_{23} \\
p_{31} & p_{32} & p_{33}
\end{bmatrix}
$$

where the probability that a number of points in subarea a_1 in time period t will 'move' to a_3 in time period $t + 1$ is p_{13}. The word 'move' can mean that the object or objects (point or points) in one subarea migrate to another subarea during the time period $t + 1$. It could be assumed, however, that 'move' means 'death in subarea a_1 and birth in a_3'. The number of points involved in the migration is a function of the movement process being studied.

Suppose that one of the points located in a_1 may move and we wish to know the probability that the point will be located in a_3 after two time periods. The transition matrix (presumably developed from theory) is as follows:

$$
t \left\{
\begin{array}{c}
a_1 \\
a_2 \\
a_3
\end{array}
\right.
\overbrace{
\begin{array}{ccc}
a_1 & a_2 & a_3 \\
\end{array}
}^{t+1}
\begin{bmatrix}
0.7 & 0.2 & 0.1 \\
0.2 & 0.6 & 0.2 \\
0.2 & 0.2 & 0.6
\end{bmatrix}
$$

Note that the probability that a point in a_1 remains in a_1 for one time period is 0.7. All possible paths the point can take are shown below together with the probability of moving between the subareas. There are three possible paths from a_1 to a_3 in two steps. The probability of the first sequence a_1 to a_1 and a_1 to a_3 is the product of the probabilities 0.7 and 0.1 (=0.07). For a_1 to a_2 to a_3 the values 0.2 and 0.2 are multiplied together (=0.04) and for

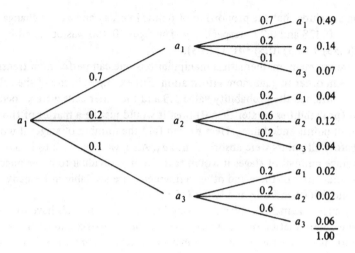

a_1, a_3, a_3 it is the probabilities 0.1 and 0.6 (=0.06). These are summed giving us the probability (0.17) that a point a_1 is found in a_3 after two stages.

We can simplify this operation by squaring the transition matrix:

$$
\begin{bmatrix} 0.7 & 0.2 & 0.1 \\ 0.2 & 0.6 & 0.2 \\ 0.2 & 0.2 & 0.6 \end{bmatrix} \cdot \begin{bmatrix} 0.7 & 0.2 & 0.1 \\ 0.2 & 0.6 & 0.2 \\ 0.2 & 0.2 & 0.6 \end{bmatrix} \begin{matrix} a_1 \\ = a_2 \\ a_3 \end{matrix} \begin{matrix} a_1 & a_2 & a_3 \\ \begin{bmatrix} 0.55 & 0.28 & 0.17 \\ 0.30 & 0.44 & 0.26 \\ 0.30 & 0.28 & 0.42 \end{bmatrix} \end{matrix}
$$

which gives all combinations of point destinations given origination locations after two stages.

For three stages we would raise the transition matrix to the third power and, in general, for n stages the transition matrix would be raised to the n th power. The matrix elements change less and less with each succeeding multiplication. The convergence of the matrices to an equilibrium state occurs after about six or seven stages.

Now let us assume at the initial stage of the system the proportion of points in $a_1, a_2,$ and a_3 is 0.6, 0.3, and 0.1, respectively. One can derive the subsequent proportions after n stages by multiplying the matrix of transition probabilities by the vector (0.6, 0.3, 0.1). Thus the state of the system after two stages would be

$$
(0.6, 0.3, 0.1) \cdot \begin{bmatrix} 0.55 & 0.28 & 0.17 \\ 0.30 & 0.44 & 0.26 \\ 0.30 & 0.28 & 0.42 \end{bmatrix} = (0.450, 0.328, 0.222)
$$

We can conclude that the proportion of points in a_1, a_2 and a_3 has changed to 0.450, 0.328 and 0.222 respectively. The figure 0.450 was obtained by (0.6) (0.55) + (0.3) (0.3) + (0.1) (0.3).

There are many mathematical manipulations one can perform on transition matrices in order to gain more information. For example, if one of the cells, say a_3, contained the probability value 1.0 and the other cells were associated with it (probabilities greater than 0) then it would just be a matter of time before all points ended in a_3. Here we can find the number of stages it would take before all points were absorbed into a_3. Also, we may wish to know the average number of stages it would take for an individual to move once. The mathematics of these and other refinements are available in Kemeny and Snell (1960) and Cox and Miller (1965).

Of course, the main problems in using Markov-chain models have to do with the determination of the transition probability matrix and with the assumptions involved in using such a matrix. If data relating the subareas in question already exist and are used to develop the matrix then the process is nothing more than a simulation. For the technique to be applied in the spirit of inquiry supported in this book, one would expect to be testing a theory of movement, the theory being embedded in the transition matrix. Unfortunately, one must assume that the transition probabilities are invariant during the temporal extent of the process. The moment it is believed that the theory is inoperative in a particular circumstance one must reconsider the nature of the transition matrix.

There is a small literature on the use of Markov processes in geography and an even smaller literature on point pattern analysis. Markov models are discussed and applied by Brown (1963, 1970), Olsson and Gale (1968), Norcliffe (1968) and Tinline (1968). Harvey (1967) makes a strong case for the use of probabilistic models in studying spatial patterns and cites Markov models as of particular interest. Brown (1963) studied the spread of an innovation (the use of propane tanks) throughout the townships in a county of Wisconsin by one year periods using a Markov model. He was tempted to use a straightforward distance—population size variable as his transition matrix building stone, but found that the matrix, if it was to predict the resulting pattern with any accuracy, had to be altered to reflect more information on the differing propensities of the people of the townships to accept the innovation. Rogers (1968) points out that Markovian concepts provide useful indices for the study of groups of potential migrants. Demographers have found that once a population is divided into its component parts, i.e. age groups, sex, race, one can seek indices of differential migration propensity. A further example of this approach can be found in Tarver and Gurley (1965).

4.1.2. Simulation

Much more popular have been the simulation models of the
Hägerstrand school of Swedish geography (see Brown (1965) and Shaw
(1975) for summaries and bibliographies on spatial diffusion and migration).
Perhaps because movement processes are so complicated, it was thought
that experimental interaction matrices could be used to replicate generally
the spatial process of the diffusion of innovation. The technique depends
on a series of assumptions concerning the way points are created. Points
represent innovators and adopters. For example, upon meeting a potential
adopter, an innovator creates an adopter who then becomes an innovator;
he too converts those that he meets. The distance separating innovators
and potential adopters determines the probability of contact. Distance
matrices are determined from migration and telephone distance data.
Random numbers are employed to determine which potential meetings
take place and maps are created for the patterns after a certain number of
rounds of pairwise meetings take place. These maps are then compared to
real-world data.

The more complicated models take into account the human resistance
to the acceptance of an innovation and the physical resistance caused by
barriers such as lakes and mountains. But it is clear that many variables having
to do with pattern change in a particular circumstance can be introduced in
order to bring more realism to the spatial process. The models of Hägerstrand
(1952, 1957, 1967) and Morrill (1963, 1965) are good examples of pattern
simulation.

Simulation models have advantages over the distributions previously mentioned in describing contagious processes; many more complicated causal
factors can be included in the analysis. The difficulty with such models,
however, is that they usually lack generality. The rules developed are usually
specific to an empirical instance. In fact it is not uncommon for the
probability distributions to be constructed from the data extracted from
the empirical situation which is to be simulated. For this reason it is very
difficult to test hypotheses based on simulation models.

4.1.3. Urn models

A third kind of 'truly contagious' model is a simulation of sorts, but
unlike a simulation an urn model leads to parametric investigations. Let us
demonstrate this point. Suppose that there is an urn which contains a number
of black beads (b) and red beads (r). A bead is drawn at random from the
urn and its color noted and returned to the urn with an additional number,
c, of beads of the color drawn. If c is positive, a reinforcement (increase in
the probability of selection) of the color drawn is experienced while a negative c implies a decline in the probability that the color drawn will be drawn
again. The parameter $p = b/(r + b)$ represents the likelihood of drawing a

black bead and q is the probability of selecting a red bead ($q = 1 - p$). The reinforcement effect is embodied in the parameter γ which is equivalent to $c/(r + b)$.

If we define x as a random variable to represent the total number of times a black bead is drawn and repeat the drawing procedure n times, we have both a simulation of an n-step process and a probability distribution. The probability model is named a Polya—Eggenberger process after its originators (see Bosch, 1963):

$$P(x; n, p, \gamma) = \binom{n}{x} \frac{\displaystyle\prod_{j=0}^{x-1} (p + j\gamma) \prod_{j=0}^{n-x-1} (q + j\gamma)}{\displaystyle\prod_{j=0}^{n-1} (1 + j\gamma)} \quad \text{for } x = 0, 1, 2, \ldots \tag{4.1}$$

where the symbol π is used for the multiplicative operation,

$$\prod_{i=1}^{n} a_i = a_1 a_2 \ldots a_n$$

In equation (4.1) when the limit $(x - 1)$ of the first numerator term or $(n - x - 1)$ of the second numerator term is less than zero, set the term equal to one. The first two moments are:

$$\mu = np \tag{4.2}$$

$$\sigma^2 = \frac{npq\,(1 + n\gamma)}{(1 + \gamma)} \tag{4.3}$$

The choice of an appropriate p and γ depends on the nature of the hypothesized process.

Spatially we can denote an urn as an area and the selection of a black bead as the placement of a point in the area. However, the model as it is fails to qualify as a spatial model, with one urn representing only one area.

Dacey (1969*b*) has shown that the process can be generalized to include S subareas. First we focus attention on a particular subarea of a region which is divided into $S(s = 1, \ldots, s^*, \ldots, S)$ subareas. If we equate an urn with s^* and the number of b black beads and r red beads with some characteristic of s^*, such as population or size of area or any one of a number of attributes, then various models of spatial processes occurring within s^* may be constructed. The meaning of γ in this context is related to the propensity of objects (points) to locate in s^*. For every black bead selected in successive drawings, an object is placed in s^*. The Polya model gives the probabilities for having x objects located in s^* after n drawings. The selection of a red bead represents the location of an object in some other subarea. After n drawings for each s, each will have a number x ($x \leqslant n$) representing the number of objects contained within it. Equation (4.1) is the model for each series of drawings, and together these make up the location process. In the situation just described we would assume that each subarea has its own attribute measure, p_s, at the start of the process. If it is assumed that γ is constant throughout the region, we have

$$P_s(x;n,\gamma) = \frac{1}{S} \sum_{s=1}^{S} P(x;p_s,n,\gamma) \qquad (4.4)$$

where P_s is the probability of observing a subarea with x objects contained within it. $P(x;p_s,n,\gamma)$ is the same as (4.1), but here the subscript s implies that there is a probability function for each subarea.

The difficulty with such a relatively simple scheme is that the diffusion takes place within subareas and not between them. One must assume that subarea boundaries are not breached. In essence the location process described is really the sum of s processes, each independent of the other. If the p_s and γ parameters are unlikely to express such a process then, of course, the model as developed is inappropriate.

The problem of dealing with impervious boundaries can be handled in the following way. Suppose it is suspected that the propensity for diffusion is a function of the location of a subarea relative to other subareas. Consider, too, that a certain group of subareas contiguous to the subarea representing the center of diffusion will be most affected by the diffusion process. If it is known or hypothesized what effect the position of one subarea has on another, then in the assignment of the p_s's and γ values the spatial relationships can be built into the model.

As an example of the use of a Polya model let us suppose that our interest is in the pattern of points in a region divided into 12 subareas. The specific question is: After 10 points are placed in the region what is the probability that a given subarea contains 0, 1, 2, ..., 10 points? It is assumed that the reinforcement effect (γ) is 0.2. Two subareas are assigned a p value of 0.8, one subarea is given $p = 0.4$, for three subareas $p = 0.2$, for another three subareas $p = 0.1$, and for the last three subareas $p = 0.0$. Using (4.4) the probabilities associated with random variables of each of 12 probability functions are summed for each of $x = 0, 1, 2, ..., 10$. Each sum is divided by 12 giving the following probabilities:

x	p
0	0.446
1	0.138
2	0.088
3	0.062
4	0.045
5	0.037
6	0.031
7	0.031
8	0.034
9	0.045
10	0.049

We can conclude that 0.446 of the 12 subareas are not expected to contain any of the ten assigned points. Apparently, the best fitting discrete set of frequencies of points would have six subareas with no points, three subareas with one point, two with two points and one subarea with three points. This latter determination was made by trial and error.

There is a modest literature in this area. Getis (1974) attempts a spatial application and Anderson (1972) shows how an urn model (but not a Polya–Eggenberger distribution) can be helpful in a simulation study. Feller (1968), Johnson and Kotz (1969) as well as Dacey (1969b) are among those who explore the properties of the Polya–Eggenberger distribution. It should be noted that one of the statistical limits of the Polya–Eggenberger distribution is the negative binomial, suggesting that these are closely related distributions.

4.2.   Disturbed lattice models

In this section, instead of beginning with a Poisson location process as we did in chapters 2 and 3, we begin with a lattice of equally spaced points representing expectations due to processes of competition and territoriality. Since in nature these equal-spaced patterns are rarely found, there have been modifications to theory calling for a 'disturbing' assumption, i.e. an assumption which states that the original lattice model will be altered in a certain way.

Theories of equal-spacing are well known. For example, in human settlement theory, service center towns are expected to be evenly spaced (central place theory). This results from competition for customers as entrepreneurs in towns extend their trade areas as far as possible, but are outcompeted at long distances by others with similar objectives. In plants, if within a species there is no great range in size, by competition the expected pattern of individuals is thought to be relatively even. There are other theories of human and animal spacing which call for such patterns.

Very often the question raised by researchers is how to account for variations from the expected even location patterns. Several general models have been proposed. One adjusts the location of each lattice element according to some preconceived notion of the level of disturbance. Another assumes that the even lattice represents an ordered environment and variations from it represent the 'random component'. In this section we will look at the disturbance approach and then discuss the notion of an ordered environment in the next section on information theory.

Let us consider a triangular lattice of points, that is, a lattice where each point is equidistant from its six nearest neighbors. Now assume that due to certain conditions each of the lattice points is displaced. The amount of displacement corresponds to the effect the conditions have on the lattice pattern. The point pattern which emerges from the disturbance is a function of the way in which each point is displaced.

One model of interest arises if each point is displaced from its original location by a like amount, the displacement being equally likely in all directions. A test of this model would be to compare the observed mean first nearest neighbor distance with the mean distance to first nearest neighbor derived from a pattern of disturbed points having the same density as the observed distribution.

The theoretical disturbed pattern may be simulated by randomly selecting an azimuth for each lattice point and 'moving' the lattice point a specified distance along the azimuth (see fig. 4.1). This procedure was followed in order to create table 4.1. Here the standard distance of one unit was used as the original distance between the nearest lattice points. A disturbance of 0.2 implies that each lattice point was moved 0.2 units along a randomly selected azimuth. The data in table 4.1 result from operations on an experimental lattice of 200 points (20 x 10) in which no measurements were taken from border lattice points, thus $N = 144$. Fig. 4.1 shows the pattern from which the values in table 4.1 for a 0.2 disturbance were derived.

For tests, it would be necessary to convert the units of the observed mean first nearest neighbor distance to the standard chosen. Lattice points each of which are spaced at one unit distance from six other points have a density of one point per 1.155 square units. If an observed lattice pattern has density d, the distance between points in map units rather than standard units would be $1.0746/\sqrt{d}$. It follows that the ratio between the observed mean first nearest neighbor distance $(\bar{r})$ and the lattice distance is the standardized distance, $\bar{r}_s$.

$$\bar{r}_s = \bar{r}/\frac{1.0746}{\sqrt{d}} \tag{4.5}$$

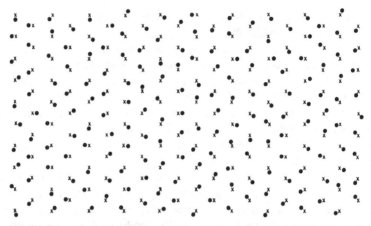

Fig. 4.1 Triangular lattice (distance 1.0 separating points marked X) disturbed by 0.2 (shown by pattern of filled circles)

Table 4.1. *Mean distance and variance of nearest neighbors in disturbed triangular lattice (sample:* N = *144): distance between original points in lattice is equal to 1.0*

Disturbance	$\hat{\mu}$	$\hat{\sigma}^2$
0.0	1.000	0.0000
0.1	0.878	0.0024
0.2	0.744	0.0099
0.3	0.656	0.0215
0.5	0.485	0.0792
Poisson model expectations		

$\mu = 0.465$
$\sigma^2 = 0.0591$

As an example suppose it is presumed that the pattern shown in fig. 2.3 is a result of a process which gives rise to an even distribution of points, but where a disturbance of 0.2 is allowed as an acceptable deviation from the uniform lattice. The density is 2 and $\bar{r}$ is 0.3061 (from table 2.4). Thus the standardized value $\bar{r}_s = 0.4028$. From table 4.1 we note that for a disturbance of 0.2 the expected mean distance to first nearest neighbor is 0.744 and the variance is 0.0099. The normal distribution may be used for the test:

$$Z = \frac{\bar{r}_s - \mu}{\sqrt{(\sigma^2/N)}} = \frac{0.4028 - 0.744}{0.0099/41} = -21.96 \tag{4.6}$$

Since we know the pattern in fig. 2.3 was in fact created by a Poisson process it is not surprising that an hypothesis of a 0.2 disturbed lattice must be rejected.

A closer look at table 4.1 shows that a disturbance of 0.5 nearly approximates a Poisson expectation thus the disturbed model only seems appropriate when small disturbances (0.1, 0.2, 0.3) are assumed. There is a danger in using a model such as this: unless the theory is sound one cannot be sure that a disturbance has in fact taken place.

Dacey and Tung (1962) and Jones (1971) describe more complicated procedures for obtaining the disturbed pattern. Instead of assuming a constant displacement distance, they consider it to be the mean of a probability distribution having a modified normal distribution. This may be useful if there is knowledge of the expected dispersion of displacement values around the mean disturbance figure. For most problems, however, it would seem difficult to make judgments about the nature of the dispersion.

Dacey and Tung (1962) were the first to introduce the disturbed lattice model. It has been described by King (1969), and Rushton (1971) has developed an algorithm for varying the lattice of points to conform to non-uniform surfaces. In this regard Tobler (1963, 1970) develops a system for evaluating distorted surfaces and Getis (1963) provides an example of point pattern responses to an heterogeneous surface.

4.3   Pattern models and information theory (quadrat approach)

Another approach to model development that also uses a uniform lattice as a base for understanding variations from it is derived from information theory. Because the theory is general, it has wider implications than the lattice-disturbance model discussed in the previous section. To date there have been relatively few applications of information theory to point pattern problems, but, with the greater interest in ideas stimulated by general systems theory, the approach is likely to become more widely used. The quadrat measurement approach is discussed here, but more useful is the area approach explained in chapter 6. The basic theory, however, is described in the following paragraphs.

The model describes the *information* elicited from a point pattern or a series of point patterns. In the context of the theory, the term 'information' has a special meaning. The placement of a point in a specific map area (quadrat) may be considered as a message generated or transmitted, but so is the intended non-placement of a point in that map area. The information associated with the placement or non-placement of a point in a specific area is called *one bit*. The information in an experiment with n equally likely outcomes is $\log_2 n$ bits. The use of logarithms is a result of the need to solve equations which are increasing functions of n. The dichotomous situation, occurrence or non-occurrence of a point, indicates that the base 2 for the logarithm is appropriate. It has become a practice to use natural logs, i.e. logs to the base e, for mathematical convenience ($\log_e = \ln$).

The information associated with an experiment having one possible outcome is $H = \ln 1 = 0$. This points to the fact that with an absolutely certain outcome, the information (H) elicited is zero. Suppose we divide a map into n different subareas. If each subarea is an equally likely message bearer then the probability that a message will be realized in subarea i is

$$p_i = \frac{1}{n} \tag{4.7}$$

Taking logarithms

$$\ln p_i = -\ln n$$

Multiply each side by $-1/n$

$$-\frac{1}{n}\ln p_i = \frac{1}{n}\ln n$$

Summing over all regions we obtain

$$\sum_{i=1}^{n} -p_i \ln p_i = \sum_{i=1}^{n} \frac{1}{n}\ln n$$

The value on the right is equivalent to $\ln n$; thus our measure of information is

$$H = \sum_{i=1}^{n} -p_i \ln p_i \tag{4.8}$$

Since

$$-\ln p_i = \ln \frac{1}{p_i}$$

we can write (4.8) as

$$H = \sum_{i=1}^{n} p_i \ln \frac{1}{p_i} \tag{4.9}$$

The information H is a measure of uncertainty. Maximum uncertainty in our context we envisage as a probability distribution where each possible outcome of a random variable has the same probability of occurrence (see fig. 4.2a). This means that it is just as likely to have a subarea with a few points in it as it is to have a subarea with many points in it. In this circumstance we have

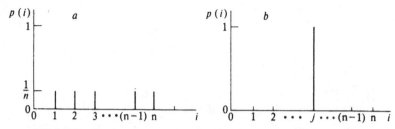

Fig. 4.2 Distributions representing (*a*) maximum and (*b*)minimum values of *H*

no way of knowing which outcome is more or less likely to occur in a subarea than any other. A 'sharply peaked' distribution, however, satisfies our intuitive notion that uncertainty is minimized and thus values of *H* will be low. As we have seen, in the extreme case when p_i equals one, *H* will be 0 (fig. 4.2*b*).

As an example of these two conditions suppose that each of a number of subareas is to receive a point. Uncertainty is minimized since the same outcome is expected for each subarea (an even point pattern). All of the probability is assigned to one outcome, thus there is one p_i and its value is 1 and *H* = 0. A greatly clustered pattern, say one having all *n* points in one subarea, also yields low *H* values since there will only be two p_i values, one of zero for the empty subareas and one for the subarea with points.

The greater the number of different p_i values for the subareas the greater the uncertainty and the higher the value of *H*. The distribution of the magnitude of p_i values affects *H*. Using (4.9), it can be seen that a p_i value of 0.360 contributes 0.368 to the value of *H* while 0.040 contributes only 0.129. In fact p_i values both higher and lower than 0.360 contribute less to *H* (see fig. 4.3). The *H* value, however, is affected more by the number of p_i values than by the value of p_i itself. This follows intuitively if we realize that the

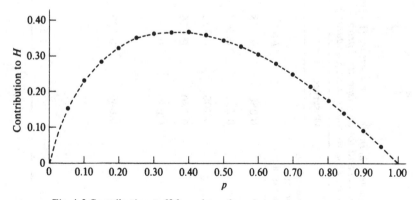

Fig. 4.3 Contribution to *H* for values of *p*

Table 4.2. *Information measure (H) for pattern shown in fig. 2.3, for Poisson process model and for an unspecified probability function approximately equal to a maximum (based on 50 points and 25 quadrats)*

Number of points per quadrat	Observed		Poisson model		Unspecified model	
	p	$p\ln(1/p)$	p	$p\ln(1/p)$	p	$p\ln(1/p)$
0	0.080	0.258	0.135	0.271	0.320	0.365
1	0.360	0.367	0.271	0.354	0.160	0.293
2	0.200	0.322	0.271	0.354	0.160	0.293
3	0.280	0.356	0.180	0.309	0.160	0.293
4	0.040	0.129	0.090	0.217	0.080	0.202
5	—	—	0.036	0.120	0.040	0.129
6	0.040	0.129	0.012	0.053	0.040	0.129
7	—	—	0.004	0.019	0.040	0.129
8	—	—	0.001	0.006	—	—
(*H*)		1.561		1.703		1.833

greater the n, the greater the possibility of more and different p_i values, the greater the uncertainty and thus the greater the value of H.

One would expect patterns generated by the assumptions of a Poisson process to yield high H values since neither even patterns nor clustered patterns could be created. To evaluate a Poisson process in terms of expected values of H, we substitute the Poisson probability function for p_i. This gives

$$H = \sum_{x=0}^{\infty} P(x; \lambda) \ln \frac{1}{P(x; \lambda)}$$

$$= \sum_{x=0}^{\infty} \frac{e^{-\lambda}\lambda^x}{x!} \ln \frac{x!}{e^{-\lambda}\lambda^x} \tag{4.10}$$

Other discrete probability distributions can be evaluated for H in the same manner.

In fig. 2.3 fifty points are located in 25 quadrats. The observed proportion for each outcome is given in table 4.2. The H value is 1.561. For a Poisson process generated pattern with $\lambda = 2$ the H value shown in table 4.2 is 1.703. The difference may be explained by the clustering of six points in one sub-area and the relatively large number of subareas with either one or three points contained within them. In table 4.2 another set of probabilities is shown, one which was constructed specifically to yield an approximate maximum H value. The reason the probabilities in the maximum distribution are not distributed as the discrete uniform (the ideal maximum) is that the fifty points in 25 quadrats greatly restricts the number of possible configurations of points in quadrats. The Poisson process value and the maximum H are not greatly different. Thus it seems a useful strategy in many cases to develop hypotheses about differences between expected and Poisson process H values.

This technique depends, of course, on quadrat sampling so that precautionary measures must be taken to ensure that an appropriate quadrat size is selected. In practice, quadrats are a poor sampling unit for many of the reasons given in previous discussions. But even more damaging is the fact that both great clustering and extreme evenness yield low H values, thus interpretation is difficult An alternative is available. In chapter 6 we discuss Thiessen polygons. This is an area concept which is used to study the region of influence of each point. These regions then become the input for the study of information content.

Quadrat studies of point patterns have been carried out by Medvedkov (1967), Semple and Golledge (1970) and Haynes and Enders (1975). Other contributors are mentioned in chapter 6.

5

Line patterns

In chapter 1 we presented a notational system which could be used to describe any of the processes we encounter in the course of this book. It was noted that point-generated processes leading to spatial patterns of lines and areas must, of necessity, contain two groups of assumptions. The requirement of two groups is warranted by the need for one group to locate the points (**P** assumptions) and a second group to locate the lines or areas (**L** or **A** assumptions) with regard to the points generated by the first group of assumptions. Although it is possible to provide a notational scheme which is capable of handling quite complex processes such as those in which both groups of assumptions change over time, research into such processes is still very much in its infancy. Consequently, most of the processes we discuss in this book are those in which the time element is either ignored or is considered irrelevant to the creation of the pattern.

In analyzing point-generated patterns our concern will be to select **P** and **L** or **A** assumptions which create an overall process which we feel might be responsible for the pattern we are interested in examining. Before doing this, however, a distinction must be made between patterns of lines and patterns of areas. Both the line and the area patterns discussed consist of two groups of components, points and lines. In *area patterns* the lines represent the boundaries of the individual areas in the pattern so that all area patterns are also automatically line patterns. The reverse is not true, for only certain kinds of line patterns can also be conceived of as area patterns. What distinguishes *line patterns* from area patterns is the nature of the processes which give rise to the patterns and the form of the lines generated by those processes. Thus, any division of processes into line-generating ones and area-generating ones must inevitably be somewhat arbitrary and artificial. However, we feel that, from the standpoint of clarity of presentation, it is worthwhile to make such a distinction.

5.1. Processes for generating line patterns

Both line and area patterns of the type we are interested in here begin with the generation of a set of points. This is achieved by any point pro-

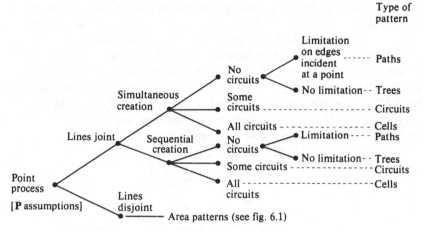

Fig. 5.1 Basic assumptions of line-generating processes

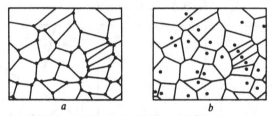

Fig. 5.2 Point and line sets. (*a*) Joint. (*b*) Disjoint

cess of the kind examined in chapters 2–4. Once the points are established the lines of the pattern are generated. Fig. 5.1 shows one way of considering patterns resulting from different groups of assumptions. First, we assume that the points and lines can be either joint or disjoint, i.e. the points may be either on or off the lines which are established (see fig. 5.2). Note that while the line characteristics of both patterns are similar in form the point patterns generated by the **P** assumptions of the processes are quite different (see fig. 5.3). We

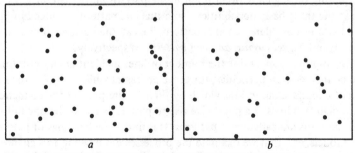

Fig. 5.3 Contrasting P assumptions of two line-generating processes

have chosen to consider line-generating processes as those leading to patterns in which the points and lines are joint, and area-generating processes as those resulting in patterns in which they are disjoint. The assumptions involved in area-generating processes will be discussed in chapter 6. As was true with point processes we may be concerned about the sequence of the development of the patterns. The lines of a pattern can be created simultaneously or sequentially. Fig. 5.4 illustrates two line patterns which are identical in final form, in one of which (fig. 5.4a) all the lines are created at the same time, while in the other (fig. 5.4b) the pattern evolves over time.

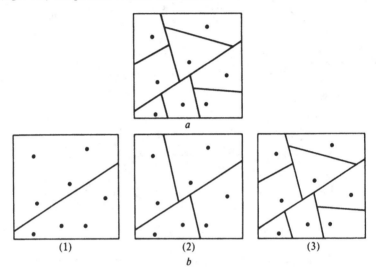

Fig. 5.4 Processes illustrating (*a*) simultaneously generated lines, and (*b*) sequentially generated lines

As fig. 5.1 shows we are also concerned about processes which entail the generation of circuits. A circuit is equivalent to a closed loop which makes it possible to trace a round trip route from at least one point in the pattern without passing through any other point in the pattern more than once. Fig. 5.5 shows the three basic possibilities. Note that the pattern produced in fig. 5.5c could also be considered an area pattern. We call the patterns shown in fig. 5.5a, b, and c tree, *circuit, and cell patterns*, respectively.

Assumptions concerning the number of lines incident at any point may also be included. In particular, for those processes which prohibit the formation of circuits, assumptions which by their nature prevent the incidence of more than two lines at a point give rise to patterns such as the one in fig. 5.6. Such patterns are called *path patterns*. It is under the headings of path models, tree models, etc. that we examine the processes discussed in this chapter.

Fig. 5.7 illustrates different patterns which result from processes involving

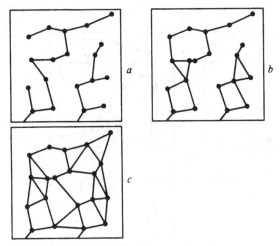

Fig. 5.5 Line patterns generated using different circuit forming assumptions. (*a*) No circuits. (*b*) Some circuits. (*c*) All circuits

different combinations of the assumptions governing the generation of lines. In fig. 5.7*a* the lines are created simultaneously and no circuits or the incidence of more than two lines at a point is permitted. The resulting pattern is a path. In fig. 5.7*b* the line components of the pattern are again produced simultaneously, but the assumption prohibiting the incidence of more than two lines at a point is changed to permit such occurrences. In figs. 5.7*c* and 5.7*d* the assumptions allow for the formation of circuits. Circuit formation is limited in fig. 5.7*c*, while in fig. 5.7*d* the assumptions state that there be circuits for all points. In fig. 5.8 the processes illustrated in fig. 5.7 are repeated, although the assumption stating that there be a simultaneous creation of elements of the line set is replaced by one allowing for their sequential generation.

It should be stressed that figs. 5.2 to 5.7 represent simple line-generating processes. In fact, using the basic assumptions outlined above, it is possible to

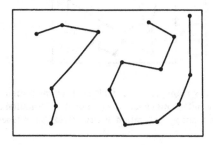

Fig. 5.6 Line pattern generated by restrictions on the number of lines incident at a point

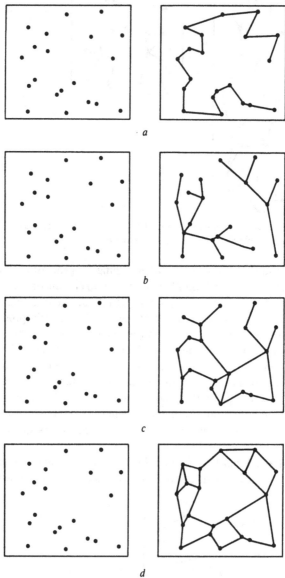

Fig. 5.7 Line patterns resulting from processes involving different combinations of basic assumptions. (*a*) Simultaneous generation of a path. (*b*) Simultaneous generation of trees. (*c*) Simultaneous generation of circuits. (*d*) Simultaneous generation of cells

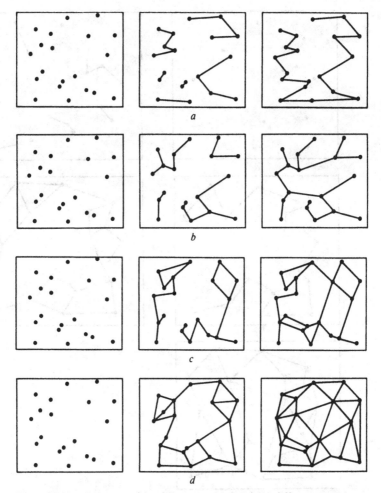

Fig. 5.8 Line patterns resulting from processes involving different combinations of basic assumptions. (*a*) Sequential generation of a path. (*b*) Sequential generation of a tree. (*c*) Sequential generation of circuits. (*d*) Sequential generation of cells

produce complex processes. In particular, if one of the assumptions permits the sequential establishment of the elements of the pattern it is possible at any time before the completion of the pattern to change any of the current assumptions governing the evolution of the pattern. This includes the **P** assumptions responsible for the creation of the pattern of points. Fig. 5.9 represents such a process. In the early stages assumptions restricting the pattern development to the formation of paths are imposed. In the later stages, once all the points have been linked to form a single path, the assumptions are changed to ones allowing for the evolution of cells.

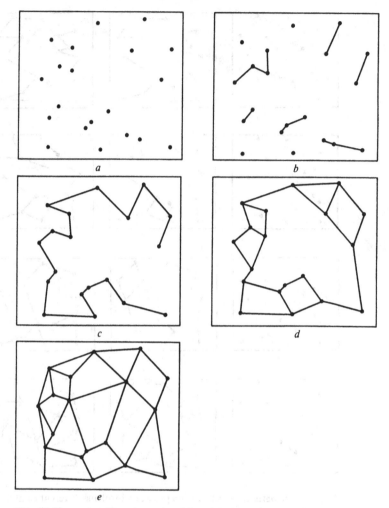

Fig. 5.9 Line-generating process involving changing assumptions

5.2.   Properties of line patterns
The line patterns which emerge from the various processes outlined above may or may not be differentiated from one another. In any case, there must be a way to describe line patterns so that judgments can be made about differences. For this we depend on the geometric properties of the patterns. These include such things as

(1)   the sum of the length of lines relative to the study area (density of lines);
(2)   the frequency distribution of the length of lines in the pattern;

(3)   the number of line intersections per unit area;
(4)   the frequency distribution of the distances between intersections;
(5)   the frequency distribution of the angles of intersection of the lines;
(6)   the number of segments in a path;
(7)   the number of circuits;
(8)   the frequency distribution of the number of edges meeting at a point;
(9)   the orientations of the lines relative to a set of coordinate axes.

Many more are possible. Clearly, the properties investigated are presumably related directly to the processes hypothesized.

To clarify the meaning of non-geometric properties we note that these include such things as the color of the points and lines, their numerical value, etc. Although we shall not discuss these it is evident that non-geometric properties add much complexity to considerations of line patterns and for many problems a great deal of realism is gained by their use. In our scheme of things the study of non-geometric properties would constitute the need for a further set of assumptions which are not of the **P** and **L** type. The models discussed subsequently may imply a need to look at non-geometric properties, but for our purposes we will limit ourselves to geometric properties.

In examining the properties of line and area patterns we encounter a feature which was not present in our discussion of point pattern properties. Points were considered to be dimensionless objects with a discrete occurrence in the study area. This meant that the number of points we observed in a pattern was always an integer, which enabled us to confine our attention to models involving only discrete probability distributions. Lines and areas, however, are one- and two-dimensional objects, respectively. Thus, some of their properties can be non-integer and continuous in form (e.g. a line may be 3.6 units in length, an area 4.2 square units in size) which requires the use of continuous probability distributions. Deriving results from models incorporating such distributions necessitates the use of integration, although throughout the book an attempt is made to minimize such activity and to present the results in as simple and readily usable form as possible.

5.3.   Path models

We begin our discussion of line models with path models. Path models are those which generate patterns of the type illustrated in fig. 5.6. Other path models are used in operations research, but these deal with tracing a path through an already defined network. We emphasize that the models described here are concerned with the generation of paths in situations in which none existed previously.

The first model we discuss generates patterns which consist exclusively of straight lines. The model comes from Horowitz (1965) who was interested in deriving properties of the length of straight-line paths across fundamental geometric shapes. Although he was motivated by non-geographic problems – in particular, the prediction of the length of the path of a gamma-ray to the wall of a nuclear reactor and the length of a sound ray in a room from one reflection to the next – his model could be useful to those interested in such problems as traffic journeys through cities, movement of pedestrians across open spaces, (Garbrecht, 1971), and flight movement and maze-running behavior of animals.

The shapes Horowitz examines are the square, rectangle, circle, cube, and sphere. Since we have consistently used a two-dimensional study area we shall disregard the last two shapes. For any given shape the **P** assumption states that each location, X on the perimeter of the shape has an equal and independent chance of being selected as the origin of a path across the shape. The **L** assumption involves selecting an angle, θ, from the interval (0 to π radians) and constructing a straight line from X, direction θ, until the line touches another point on the perimeter of the shape, thus creating a path. Fig. 5.10 illustrates two paths generated in this manner across a rectangle of width W and height H.

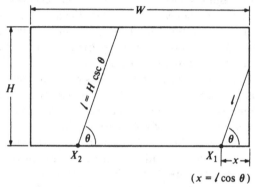

$$(x = l \cos \theta)$$

Fig. 5.10 Two different paths across a rectangle. Source: Horowitz (1965, fig. 1, p. 170)

Using the same assumptions for each of his geometric shapes, Horowitz derives both the probability density function **P** (l) of the length of a path across the shape, and $\bar{l}$, the average path length. For a *rectangle* of sides length W and $H (H < W)$

$$P(l; W, H) = \begin{cases} \dfrac{4}{\pi(W+H)} & \text{for } 0 < l < H \\[2ex] \dfrac{2}{\pi(W+H)}\left[\dfrac{WH}{l\sqrt{(l^2-H^2)}} - \dfrac{\sqrt{(l^2-H^2)}}{l} + 1\right] \\[1ex] & \text{for } H < l < W \\[2ex] \dfrac{2}{\pi(W+H)}\left[\dfrac{WH}{l\sqrt{(l^2-H^2)}} - \dfrac{\sqrt{(l^2-H^2)}}{l} + \dfrac{WH}{l\sqrt{(l^2-W^2)}} \right. \\[2ex] \left. \quad - \dfrac{\sqrt{(l^2-W^2)}}{l}\right] & \text{for } W < l < \sqrt{(W^2+H^2)} \end{cases}$$

$$(5.1)$$

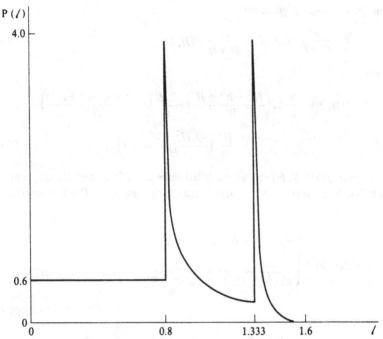

Fig. 5.11 Probability density function for the lengths of paths across a rectangle (height = 0.800, width = 1.333). Source: Horowitz (1965, fig. 4, p. 174)

In fig. 5.11 the distribution of $P(\ell; W, H)$ is plotted for a rectangle in which $W = 1.333$ and $H = 0.800$. The bimodal nature of the distribution reflects the influence of points located at the vertices of the shape. By integrating, we get the following results.

$$F(\ell; H-0) = \left[\frac{4\ell}{\pi(W+H)} \right]_0^H$$

$$F(\ell; W-H) = \frac{2}{\pi(W+H)} \left\{ WH\left(\frac{1}{H}\sec^{-1}\frac{\ell}{H} \right) - \left[\sqrt{(\ell^2 - H^2)} \right.\right.$$
$$\left.\left. - H^2\left(\frac{1}{H}\sec^{-1}\frac{\ell}{H} \right) \right] + \ell \right\}_H^W$$

$$F(\ell; \sqrt{(W^2 + H^2)} - W) = \frac{2}{\pi(W+H)} \left\{ WH\left(\frac{1}{H}\sec^{-1}\frac{\ell}{H} \right) - \left[\sqrt{(\ell^2 - H^2)} \right.\right.$$
$$\left. - H^2\left(\frac{1}{H}\sec^{-1}\frac{\ell}{H} \right) \right] + WH\left(\frac{1}{W}\sec^{-1}\frac{\ell}{W} \right)$$
$$\left. - \left[\sqrt{(\ell^2 - W^2)} - W^2\left(\frac{1}{W}\sec^{-1}\frac{\ell}{W} \right) \right] \right\}_W^{\sqrt{(W^2 + H^2)}} \quad (5.2)$$

$\bar{\ell}$ for the *rectangle* is given by

$$\bar{\ell} = \frac{W}{W + H} f(W, H) + \frac{H}{W + H} f(H, W) \tag{5.3}$$

where

$$f(W, H) = \frac{W}{\pi} \ln\left(\frac{H^2 + W^2 + H}{W}\right) + \frac{2H}{\pi} \ln\left(\frac{W + \sqrt{(W^2 + H^2)}}{H}\right)$$
$$- \frac{H^2}{\pi}\left(\frac{\sqrt{(H^2 + W^2)}}{H} - 1\right) \tag{5.4}$$

For a *square*, $P(\ell; W, H)$ can be modified since $W = H = S$, say. Consequently, when $P(\ell; S)$ is plotted for a square there is only one peak. The appropriate equations are

$$P(\ell; S) = \begin{cases} \dfrac{2}{\pi S} & \text{for } 0 < \ell < S \\[2ex] \dfrac{4S}{\pi\ell\sqrt{(\ell^2 - S^2)}} - \dfrac{2}{\pi S\sqrt{(\ell^2 - S^2)}} & \text{for } S < \ell < \sqrt{(2S)} \end{cases}$$

$$\tag{5.5}$$

By integrating we get

$$F(\ell; S - 0) = \left[\frac{2\ell}{\pi S}\right]_0^S \tag{5.6}$$

$$F(\ell; \sqrt{(2S)} - S) = \left[\frac{4S}{\pi}\left(\frac{1}{S}\sec^{-1}\frac{\ell}{S}\right) - \frac{2}{\pi S}\left(\cosh^{-1}\frac{\ell}{S}\right)\right]_S^{\sqrt{(2S)}}$$

$\bar{\ell}$ for the *square* is

$$\bar{\ell} = \frac{3S}{\pi}\left[\ln(1 + \sqrt{2}) + 1 - \sqrt{2}\right] \tag{5.7}$$

For a *circle* (radius $= R$)

$$P(\ell; R) = \frac{2}{\pi\sqrt{(4R^2 - \ell^2)}} \tag{5.8}$$

By integrating we get

$$F(\ell; 2R - 0) = \frac{2}{\pi}\left[\sin^{-1}\frac{\ell}{2R}\right]_0^{2R} \tag{5.9}$$

$\bar{\ell}$ for the *circle* is

$$\bar{\ell} = \frac{4R}{\pi} \tag{5.10}$$

As an example of an application of Horowitz's model for one geometric shape, consider fig. 5.12. It represents a circular plaza to which access and exit can be gained at any point on its perimeter. We feel that if pedestrians cross the plaza along straight-line paths and show no preference for any particular point at which they enter or leave the plaza the frequency of path lengths should be similar to that expected from the model. In order to test our hypothesis on pedestrian movement we have noted where a sample of 100 people entered and left the plaza. Some of these paths are shown in fig. 5.12.

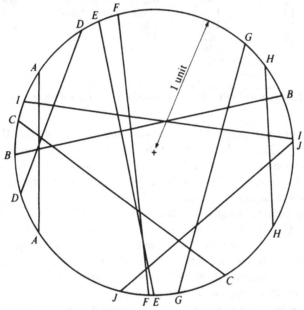

Fig. 5.12 Selected paths across a circular plaza (the letters represent individual pedestrians)

In particular, we have measured the length of each person's route through the plaza and constructed the frequency distribution given in table 5.1. For convenience we consider the radius of the plaza to be 1 unit. The empirical frequencies can be compared with those expected from the model and the degree of fit measured using a Kolmogorov–Smirnov (K–S) one-sample test. The K–S one-sample test can be used when observed values are compared to expected values taken from continuous distributions such as those developed from the Horowitz model. It was not appropriate to use the K–S test for the discrete distributions of chapters 2 and 3. One simply finds the largest absolute difference in proportions taken from a cumulative listing of the observed and expected proportions (see table 5.1). The largest difference (D) is compared to a K–S tabled value (see appendix B) for a particular critical region and an

Table 5.1. *Frequency distribution of path lengths*

ℓ	Observed frequency	Observed proportion	Observed cumulative proportion	Expected proportion	Expected cumulative proportion
0–0.20	8	0.080	0.080	0.064	0.064
0.21–0.40	8	0.080	0.160	0.064	0.128
0.41–0.60	7	0.070	0.230	0.066	0.194
0.61–0.80	6	0.060	0.290	0.068	0.262
0.81–1.00	6	0.060	0.350	0.071	0.333
1.01–1.20	9	0.090	0.440	0.076	0.409
1.21–1.40	12	0.120	0.560	0.084	0.493
1.41–1.60	10	0.100	0.660	0.096	0.589*
1.61–1.80	10	0.100	0.760	0.116	0.705
1.81–2.00	24	0.240	1.000	0.295	1.000
* $D = 0.071$	$D_{0.20} = 0.107$				

D is the calculated K–S value obtained from the sample.
$D_{0.20}$ is the K–S critical value for 0.20 level of significance given sample size.

N. The N is the total number of observations. In table 5.1 note that D is 0.071 while for $N = 100$ the tabled D value at the 0.20 level of significance is 0.107. Since the observed value is lower than the expected we conclude that the two sets of frequencies are not significantly different. (For further discussion of the K–S test see Siegel (1956) and Walsh (1963).)

To get the expected frequency distribution we use (5.9). The maximum path length will be 2 units which represents a path across the diameter of the plaza, while the minimum value is 0 which is equivalent to stepping on the plaza at one point and immediately stepping off again at the same point. The expected frequencies derived from (5.9) appear in table 5.1.

Little other work has been done on general path models although related research is involved in the examination of the expected distance of a set of points. The expected distance is an estimate of the most likely distance (measured by means of a straight line) between any two points in a pattern within a closed space. Much of this work has been reviewed by Massam (1974; 1975, pp. 96–102). There have been some applications of this work particularly in providing approximate solutions to location–allocation problems in which the precise locations of the destinations are unknown (Christofides and Eilon, 1969; Eilon *et al.*, 1970; Cooper, 1974). Garwood and Holroyd (1966) applied a similar approach in the examination of traffic flows in cities. In their analysis, cities were considered circular in shape and every point within the city was considered equally likely as either an origin or a destination of a trip, while all journeys were assumed to follow straight line paths. Other ways of generating lines across a circle have also been pursued by statisticians. In particular, the paradox usually ascribed to Bertrand deals with the probability that the length of a path drawn across a unit circle is longer than the side of an inscribed equilaterial triangle (i.e. $P(\ell) > \sqrt{3}$). The paradox arises because at least three different solutions can be presented, each of which is associated with a different way of generating the path (see Kendall and Moran, 1963, pp. 9–10). In addition, geographers Bunge (1966) and Taylor (1971) have examined the lengths of paths between all possible pairs of points within particular shapes in attempts to find suitable techniques for use in shape analysis.

5.4. Tree models

Tree models result from processes incorporating an **L** assumption which prohibits the formation of loops in the patterns generated. This means that for n points, $n - 1$ lines (edges) are required to construct a tree, and it is assumed that each pair of points is connected by one and only one sequence of edges (see fig. 5.5a for an example). For a set of n points it is possible to draw $n^{(n-2)}$ different trees. One of these, the *minimal tree*, is useful as a standard by which to compare other trees. A minimal tree is formed when an additional **L** assumption states that the aggregate length of the edges of the tree is a minimum. Fig. 5.13 shows how such a tree can be constructed. We begin with the points generated by the **P** assumptions. In this case there are twelve points located according to a Poisson process (fig. 5.13a). First each point is connected to its nearest neighbor (see fig. 5.13b). After doing this we are left with a set of disconnected elements (in this case three). We now link each disconnected element to its nearest neighboring element. This reduces the number of disconnected elements. The nearest neighbor linkage procedure is repeated until only one element, linking all the original points, remains (see fig. 5.13c). This is the minimal tree (see Scott, 1971, for a full discussion).

Minimal trees can be used as standards to judge the form of trees constructed for transportation networks, communication systems etc., especially when the prime concern is the minimization of the total length of edges in the system. Although it is not difficult to construct a minimal tree for any set of points it can become a tedious and time-consuming exercise especially when the number of points is large. Gilbert (1965) has attempted to determine whether it is possible to derive any of the properties of a minimal tree without actually constructing it. In particular, he considered the expected length of a minimal tree, $E(\ell)$, for n points located according to the **P** assumptions of a Poisson process. When the points are generated in this way, Gilbert showed that as the number of points increases (i.e. as $n \to \infty$),

$$E(\ell) \to k\sqrt{(An)} \qquad (5.11)$$

where A is the size of the study area and k is a constant. Gilbert estimated the value of k to be 0.68. Subsequent work by Roberts (1968) analytically established the bounds of k as 0.500 to 0.707, estimating the value of k as 0.656. Substituting Roberts' value in (5.11) we may estimate $E(\ell)$ for fig.

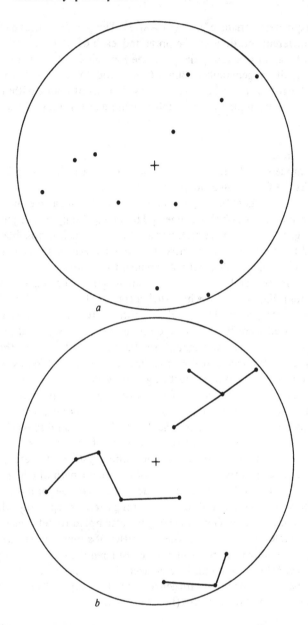

5.13. This gives $E(\ell) = 0.656\sqrt{(\pi 12)} = 4.03$. The observed length of the tree in fig. 5.13 is 3.88. Differences may be due to the small number of points in the pattern.

Gilbert's study of the length of a minimal tree is of additional interest in that he examines the *exodic tree* which is closely related to the minimal one.

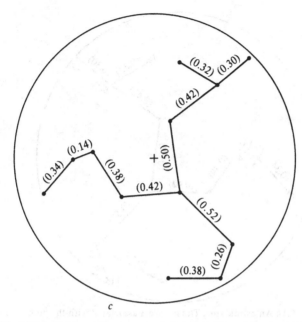

Fig. 5.13 A tree-generating process. Interpoint distances in parentheses

This 'not quite minimal tree' is one in which certain lines radiate outward from a selected central point. Suppose n points have been located in an area according to Poisson process assumptions. First, the point which is closest to the center of the study area is found, then the n points are labeled $P_1, P_2, \ldots, P_n$ so that the subscripts reflect the order of the distances the points are from the center of the area. The **L** assumption states that for $i = 2, 3, \ldots, n$, point P_i is connected to another point P_j ($i \neq j$) chosen from $P_1, \ldots, P_{i-1}$, so that the absolute distance $|P_i - P_j|$ is a minimum. The length of the exodic tree formed is the sum of all the lines $|P_i - P_j|$. The exodic tree from the same set of points used to generate the minimal tree in fig. 5.13 is shown in fig. 5.14. Since P_1 is the most 'central' point in the area Gilbert suggests that the length of an exodic tree might be expected to only slightly exceed that of a minimal tree. This was, in fact, Gilbert's primary motive for considering the exodic tree. Note that the exodic tree in fig. 5.14 differs from the minimal tree of fig. 5.13 in only one line. In fig. 5.13 point 10 is linked to point 11, in fig. 5.14 point 10 and point 8 are linked. Gilbert considers the expected length, $E(n)$, of an exodic tree drawn from a set of n points located in a unit circle according to the **P** assumptions of the Poisson process and finds that as $n \to \infty$,

$$E(n) \to \sqrt{(\pi n/2)} \tag{5.12}$$

Using (5.12) the length of the exodic tree in fig. 5.14 is expected to be

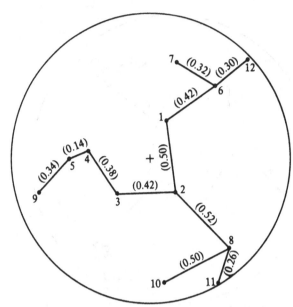

Fig. 5.14 An exodic tree. The numbers associated with the points reflect the order of the distances the points are from the center of the area. Interpoint distances are in parentheses

4.34 units. The actual length is 4.00. We should note that so far we have measured the distance between the points in the tree according to the Euclidean metric (i.e. for two points, P_i, P_j, with coordinates (x_i, y_i) and (x_j, y_j), respectively, the distance separating them is $[(x_i - x_j)^2 + (y_i - y_j)^2]^{\frac{1}{2}}$), but for some problems (e.g. distances between points on a transportation network in a city built on a grid plan) the Manhattan (right angle) metric may be more appropriate. In this case the distance between P_i, P_j is given by the expression $|x_i - x_j| + |y_i - y_j|$. For the tree in fig. 5.14 the distance according to a Manhattan metric is 5.32. Gilbert shows that as $n \to \infty$, the difference between $E(n)$ for the Euclidean and Manhattan metrics becomes insignificant.

Like the minimal tree, the exodic tree can also be used as a standard in the evaluation of communications networks, especially where some centrally located node is given a differential weighting. This would be true of communications systems which focus on specific points. Tinkler (1972, p. 6) suggests that in most cases 'geographic networks can be thought of as radial or fanning structures polarized onto a single primary node' and Haggett (1967) has already presented an instance of the reduction of a circuit network (composed of roads) into a tree form where the focus of the tree is a regional center (Viseu, Portugal; see fig. 5.15).

Finally, we should note that other tree models have been discussed in the examination of drainage patterns. However, the usual approach in this field

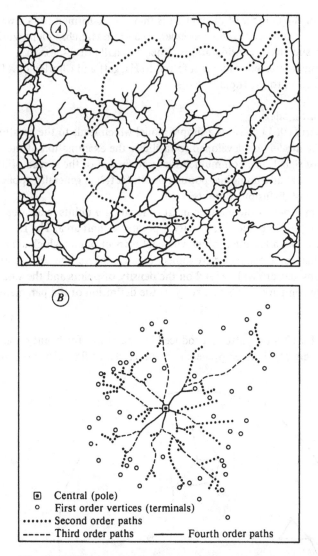

Fig. 5.15 Reduction of a circuit network to tree form. Source: Chorley and Haggett (1967, fig. 15.28, p. 660)

differs from ours and emphasizes topological properties of the trees. In such studies the locations of the points are unimportant, the only relevant property of them is their number. In addition, examination has usually been confined to trees for which additional **L** assumptions are required. The most important of these additional assumptions is that each point is the end point of one or three lines only. Much of the fundamental work in this field is based on the examination of the number of distinct trees that can be formed for a given number of points subject to these extended assumptions. Subsequent work is predicated on such basic considerations. Since we shall not discuss such models here the interested reader is referred to Haggett and Chorley (1969) for an introduction to the topic.

5.5.   Circuit models

Gilbert (1961) presents a simple model which leads to the creation of circuits. The model was developed to facilitate the examination of the transmission of signals in a communications network or of the spread of disease through a spatially immobile population. In fact, it is generally applicable to a wide range of diffusion contexts.

The **P** assumptions of the model consist of locating n points in an area according to a Poisson process of density λ points per unit area. The **L** assumption involves a linking process where all pairs of points which are separated by less than some distance r are joined. Clearly, the number of circuits produced in any pattern will depend on the density of points and the value of r. To facilitate comparison, Gilbert suggests the definition of the parameter

$$E = \pi r^2 \lambda \tag{5.13}$$

Figs. 5.16 and 5.17 show patterns produced by the model for twenty points located according to a Poisson process in a unit circle. In fig. 5.16, $r = 0.38$, so

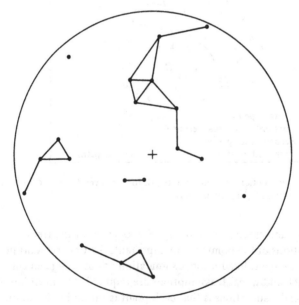

Fig. 5.16 A circuit-generating process ($E = 2.89$)

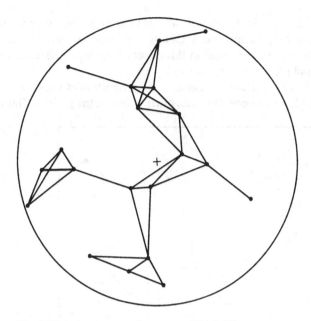

Fig. 5.17 A circuit-generating process (E = 5.41)

that E = 2.89, while in fig. 5.17, r = 0.52 and E = 5.41. The difference in the two patterns is obvious. In particular, in fig. 5.17 none of the points is isolated and they are all linked into a single structure. Thus, for our pattern there must be some value of E between 2.89 and 5.41 at which all of the points become linked into a single structure. One of Gilbert's concerns was to estimate this critical value of E. In addition, he attempted to obtain the probability, $P(n)$, that a point is linked to at least $n - 1$ other points of the pattern. Although Gilbert was able to establish upper and lower bounds for E, he was unable to determine analytically the exact values of $P(n)$ and he resorted to estimating $P(n)$ from a large number of computer-drawn patterns. In fact, as we shall see in chapter 8 (pp. 152–6), the same problem was subsequently studied by Roach (1968) in the context of area patterns. Roach was able to derive analytically a good approximation for $P(n)$. This is

$$P(n) \simeq [1 - \exp(-\pi r^2 \lambda)]^{n-1} [\exp(-\pi r^2 \lambda)] \tag{5.14}$$

From his computer-drawn patterns Gilbert estimates the critical value of E to be 3.2. This yields a value of $r = \sqrt{(1.019/\lambda)}$. More recent work by Roberts (1967), again using estimates derived from computer-drawn patterns, suggests that Gilbert's estimate of E may be too small. Roberts gives a critical value of E = 3.82, so that $r = \sqrt{(1.216/\lambda)}$. Since both researchers estimate their values from actual patterns we cannot judge which value of E is more appropriate. However, we may note that for the pattern of points in figs. 5.16 and 5.17

the critical value of E is 5.04, although we must remember that the pattern shown consists of only twenty points. Clearly more work is needed in this area of research. Some recent work in this direction has been undertaken by Ling (1973*b*) and Naus and Rabinowitz (1975).

Expression (5.14) enables us to estimate the proportion of points in a pattern that will be linked to one, two, etc. other points in the pattern. This can

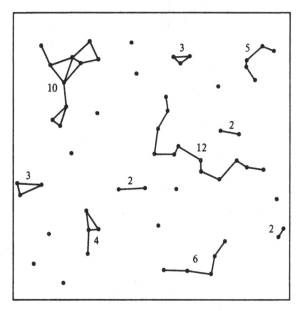

Fig. 5.18 Circuits produced by interaction of itinerant merchants. The numbers represent the size of the market system

be useful to us in pattern analysis. Suppose that the points in fig. 5.18 represent the home bases of itinerant merchants serving a dispersed agrarian society and that each merchant has a maximum range of 5 kilometers from his base. Further, we assume that merchants both sell their own goods and also occasionally buy the goods of other merchants for later resale. In this way the goods produced by one merchant may appear outside the marketing range of that particular merchant. By examining the dispersion of goods over the area we are able to identify distinct marketing systems and the number of merchants involved in each one. These systems are identified in fig. 5.18 and summarized in table 5.2. The process involved in the marketing structure is similar to that incorporated in the model if we assume that the home bases of the merchants are located following a Poisson process. If so, we can use expression (5.14) to predict the size frequency distribution of the marketing systems. We can test the fit of this expected data with the observed data using a X^2 test. The procedure is illustrated in table 5.2. The χ^2 critical value for the 0.20 level of

Table 5.2. *Size frequency distribution of marketing systems*

Size of marketing system (*n*)	No. of merchants in groups of size *n*	
	Observed	Expected
1	11	9.108
2	6	7.728
3	6	6.552
4	4	5.556
5	5	4.716 ⎫
6	6	3.996 ⎭
≥7	22	22.344
(*N*)	60	60

$X^2 = 1.868$ $df = 4$ $\chi^2_{0.20} = 5.989$

significance is 5.989 for $4df$ while the calculated X^2 value obtained from the data is 1.868. We may conclude that the model is effective in describing mercantile behavior in this instance.

Brown (1965, 1968) has examined Gilbert's model in a general diffusion context, again generating selected patterns using a computer. Haggett and Chorley (1969) have suggested that the model may also be used as a standard by which to evaluate transportation patterns of a circuit type.

5.6. Cell models

The processes involved here generate patterns which consist exclusively of closed loops. Such patterns also have the property of exhaustively dividing up the plane into a set of contiguous areas. Thus, cell models give rise to two kinds of patterns, one of lines, and one of areas, which may be considered the geometric inverses of each other. We could examine such models either in the

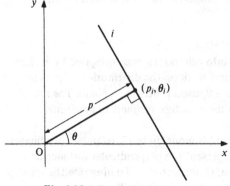

Fig. 5.19 A line defined in terms of polar coordinates

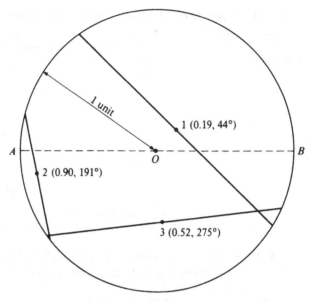

Fig. 5.20 A cell-generating process. Figures in parentheses are (p_i, θ_i)

context of line patterns or area patterns, or both. Here we choose lines because the processes discussed primarily involve locating lines in the plane.
There is no conscious effort to generate areas, and the creation of such features
is incidental to the process. When working empirically with such models the
choice of which set of features to examine is in the hands of the researcher.
Obviously, his decision will be influenced by the nature of his particular
empirical circumstances. Often, however, the two properties of line and area
patterns cannot be meaningfully examined in isolation since in cellular networks the lines, or line segments, necessarily form portions of the perimeters
of the areas in the pattern. Although the processes we describe emphasize the
generation of lines, we shall consider properties of the line patterns and of the
associated area patterns.

5.6.1.   The Miles model

Most of the early research into cell models was instigated by R. E.
Miles as part of his extensive analytical work on the distribution of points,
lines, and areas located according to a Poisson process in a plane. The model
we discuss was originally applied to the structure of fibers in paper-making
(Miles, 1964a, 1964b).

Miles chooses to define the lines in his model in terms of polar coordinates
(p_i, θ_i) (see fig. 5.19), where p_i, θ_i represent the perpendicular distance and
direction of a line i from some origin, O, respectively. To illustrate the assumptions Miles uses to generate his lines consider fig. 5.20. We begin by drawing

the area in which the lines are to be located. In this case it is a unit circle. The center of the circle is selected as the origin of the polar coordinate system through which we draw an initial reference line *AB* (corresponding to the *x* axis in fig. 5.19). The **P** assumptions of the model state that pairs of polar coordinates (p, θ) are located (selected) according to a Poisson process. Values for *p* can be chosen in the range (0 to *r*) corresponding to the radius of the

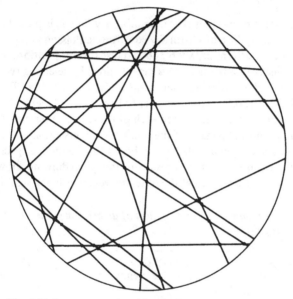

Fig. 5.21 A pattern produced by a cell-generating process

circle, and values of θ can be chosen in the interval (0 to 2π; recall 2π radians equals 360°). The **L** assumption states that a line through each of the points (p_i, θ_i) is selected so that the line produced is perpendicular to one joining (p_i, θ_i) to the origin. Fig. 5.20 demonstrates this procedure for three lines, while fig. 5.21 shows a pattern created by drawing twenty lines in this fashion. For convenience we shall designate the pattern of lines formed, $\mathcal{L}_M$. The advantage of this approach is that a uniform density of lines is maintained throughout the area. We should note that the location of *O* is not significant and that $\mathcal{L}_M$ can be generated using any point in the area as the origin. For the pattern in fig. 5.21, define λ as the density of points p_i in the interval (0 to *r*). Since the area is a circle, p_i can have values in the range $-1.0 \leqslant r \leqslant 1.0$, so that the absolute length of the interval is 2.0 units. Thus, $\lambda = 20/2 = 10.0$. In fact, because of the nature of the **P** assumptions of the model the distances ... $p_{-2} \leqslant p_{-1} \leqslant p_0 \leqslant p_1 \leqslant p_2 \ldots$ of the lines constitute the coordinates of a Poisson process on a one-dimensional axis of constant density λ.

Miles shows that in a pattern generated by this model the expected length of a line $E(\ell)$ per unit area will be equal to λ, so that the total length of the lines, L_A, in an area of size A will be $A\lambda$. For our pattern (fig. 5.21) this gives values of $E(\ell) = 10.0$ and $L_A = \pi\lambda = 31.4159$. The observed values are 9.965 and 31.306, respectively. In addition, Miles states that the number of vertices per unit area, V, which is also the number of line intersections, is equal to λ^2/π. This gives an expected value for the pattern of 31.831; the observed value is $96/\pi = 30.558$.

A second important property is that if an arbitrary line, ℓ, is placed over $\mathcal{L}_M$, the points of intersection of ℓ with $\mathcal{L}_M$ constitute a Poisson process of density $2\lambda/\pi$. For our pattern, $2\pi/\lambda = 6.366$. To test this proposition, select twenty arbitrary lines of unit length. This may be done by selecting twenty values of θ from a random numbers table in the interval $0 \leqslant \theta < 2\pi$. These values can be used as the orientations of twenty lines of unit length with one end at O. We can then observe how often each of these lines intersects $\mathcal{L}_M$. A frequency distribution can be made of the number of intersections and, using a X^2 test, this can be compared with the number of intersections expected for a Poisson process of density $2\lambda/\pi$. The results of this procedure are shown in table 5.3. As we see we cannot reject the observed values as being significantly

Table 5.3. *Frequency of intersection of an arbitrary line of unit length with $\mathcal{L}_M$.*

Number of intersections	Observed frequency	Expected frequency
0	0	0.034 ⎫
1	0	0.218 ⎪
2	2	0.696 ⎬
3	0	1.478 ⎪
4	3	2.352 ⎪
5	3	2.996 ⎭
6	2	3.178 ⎫
7	4	2.890 ⎬
8	6	2.300 ⎫
9	0	1.626 ⎪
10	0	1.036 ⎬
$\geqslant 11$	0	1.196 ⎭
(N)	20	20

$X^2 = 0.012$ $df = 1$ $\chi^2_{0.20} = 1.642$

different from those expected from the model since the χ^2 value for the 0.20 level of significance is 1.642 for 1 df while the X^2 value obtained from the sample is 0.012. We should also note that each of the lines in $\mathcal{L}_M$ can be considered as an arbitrary line and we see that the mean number of intersections per unit length = 19/31.306 = 6.069 which is close to the predicted value of 6.366.

In addition, Miles provides the expected values for properties of the polygons which result from the process. In particular, he gives first, second, and

Table 5.4. *Moments for properties of polygons in a pattern generated by the Miles model and a comparison with some values obtained from fig. 5.21*

Property	Expected value	Observed value
Number of sides		
$E(N)$	4.00	3.93
$E(N^2)$	$(\pi^2 + 24)/2 = 16.93$	16.28
Perimeter		
$E(S)$	$2\pi/\lambda = 6.28\lambda^{-1}$	$5.74\lambda^{-1}$
$E(S^2)$	$\pi^2(\pi^2 + 4)/2\lambda^2 = 68.44\lambda^{-2}$	$50.98\lambda^{-2}$
Length of any side		
$E(L)$	$\pi/2\lambda = 1.57\lambda^{-1}$	$1.46\lambda^{-1}$
Area		
$E(A)$	$\pi/\lambda^2 = 3.14\lambda^{-2}$	$2.26\lambda^{-2}$
$E(A^2)$	$\pi^4/2\lambda^4 = 48.70\lambda^{-4}$	$18.63\lambda^{-4}$
$E(A^3)$	$4\pi^7/7\lambda^6 = 1725.87\lambda^{-6}$	$198.48\lambda^{-6}$

λ = density of points in the interval $(0, r)$.
Source: Miles (1964a).

third order moments of the areas of the polygons, $E(A), E(A^2), E(A^3)$, the first and second order moments of the perimeter lengths of the polygons, $E(S), E(S^2)$, and the first and second order moments of the number of sides of the polygons, $E(N), E(N^2)$. These appear in table 5.4. Also in the table for comparison are values calculated from fig. 5.21. In computing the observed values we exclude any polygons which have a portion of the study area as one of their edges.

Miles also shows that the distribution of perimeter lengths $2\lambda S/\pi$ for polygons of k sides ($k = 3, 4, 5, \ldots$) is distributed as χ^2 with $2(k - 2)$ degrees of freedom. This means that the expected length, $E(S)$, of a 3-sided polygon is π/λ and the mean length for a class of k-sided polygons is $[(k - 2)\pi]/k\lambda$. In addition, the probability that a polygon is a triangle is $[2 - (\pi^2/6)]$ or 0.3551. The proportion of triangles in fig. 5.21 is 0.3600, while their mean perimeter length, $E(S)$, is 0.2745 compared with the expected value of 0.3142.

Santalo and Yanez (1972) have also examined the same process and their results correspond to those of Miles (1964a). Although we do not expect to encounter real-world patterns which look exactly like those produced by the model we may encounter patterns which are similar to model-generated ones in certain aspects (e.g. the distribution of the sizes of the areas). Consequently, the model can be used as a standard by which to evaluate and compare real-world patterns, in a role somewhat analogous to that sometimes played by the Poisson distribution in relation to point patterns. However, we should note that the analogy is not quite perfect. Patterns of points generated by Poisson processes occur in arnry situations, while it is difficult to think of any natural processes which produce line patterns of the type formed by $\mathcal{L}_M$.

5.6.2. The Dacey tests

Dacey (1963b, 1967) has provided a series of tests which can be used to evaluate the Miles model. All are based on the intersection of a sample line with the lines of the observed pattern. Dacey's primary concern is with patterns in which each unit of the study area has an equal probability of receiving an ω or part of an ω (where ω is either a point, line, or surface). The basic Miles model produces a pattern which meets Dacey's requirements. In addition, the techniques can also be used in the analysis of any line pattern, $\mathcal{L}$, in which

   (i)   the lines are straight or arced;
   (ii) the lines meander but do not frequently reverse direction; and
   (iii) the observed mean length of lines is at least as long as $1.5 \, E(r_i)$
         (where $E(r_j)$ is the expected distance to the jth nearest neighbor). He
         suggests that the theoretical value may be $(\pi/2)E(r_j)$. (Dacey, 1963b,
         p. 541).

Nearest neighbor approach. The technique involves selecting points from within $\mathcal{L}_M$, or more generally $\mathcal{L}$, and analyzing these using a nearest neighbor technique (see chapter 2, pp. 26–32). The pattern is described by examining the mean perpendicular distance from randomly selected points on lines of $\mathcal{L}$ to each of the first, second, ..., j^{th} nearest neighboring lines. Let us examine the approach more closely. First label all distinct lines in $\mathcal{L}$, $L_1, L_2, \ldots, L_n$. Now define i, a point on any line of $\mathcal{L}$. If we examine fig. 5.22 we see that the distance from i on L_i to L_j is the distance from i to a point j on L_j that is closest to i. This is the same as the perpendicular distance with respect to L_j. In this way the first, second, ..., j^{th} closest lines to i can be identified. Having defined a method for distinguishing the jth nearest neighboring line of i, we now need a method of drawing a random sample of n such points i. To achieve this Dacey suggests two practical ways:

   (i)   divide the lines of $\mathcal{L}$ into equal lengths and assign consecutive
         numbers to the midpoints of the segments. The sample points i are
         selected by drawing at random from within this set of midpoint
         numbers;

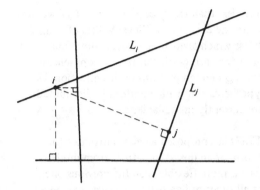

Fig. 5.22 Definition of nearest neighboring lines

(ii)   select a point i at random on each of the lines of $\mathcal{L}$. The complete
set of such points forms the sample.

Dacey shows that for line patterns in the plane the mean distance to the jth
nearest neighbor, $E(r_j)$, is given by

$$E(r_j) = \frac{0.31831j}{\hat{\lambda}} \qquad (5.15)$$

while the variance, $v(r_j)$, is

$$v(r_j) = \frac{0.10132j}{\hat{\lambda}^2} \qquad (5.16)$$

where $\hat{\lambda}$ is the estimate of λ, the density of lines per unit area of the pattern,
estimated from the observed density, d, i.e. $\hat{\lambda} = d$, where

$$d = \frac{\text{total length of lines}}{\text{size of study area}} \qquad (5.17)$$

However, if the number of lines in the study area is small and we use the
second method of drawing a sample, Dacey suggests that

$$\hat{\lambda} = \frac{d(L-1)}{L} \qquad (5.18)$$

where L is the number of lines in the study area, is a better estimate of λ.

For those interested in a demonstration of this technique see appendix A
of this chapter.

If the differences between observed and expected values are significant the
possible interpretations are similar to those presented when the nearest neighbor technique was explained in chapter 2. If the hypothesis that the pattern
could have been the result of the model process is rejected one must conclude
that the pattern may be the result of another process. If the mean of the ob-

served values, r_j obs, is significantly less than the expected value, $E(r_j)$, we most likely have a process which leads to the clustering of lines. Values of r_j obs greater than $E(r_j)$ indicate a process which leads to lines being more uniformly distributed. In fact, Dacey suggests that this particular test is more sensitive in detecting departures from model-generated patterns in this direction. Unfortunately, rigorous tests of hypotheses concerning uniformly distributed lines or the clustering of lines are currently unavailable.

Reciprocals approach. Dacey (1967) has proposed an alternative test for examining the departure of the form of an observed pattern from model-generated patterns. This technique is more flexible than the previous one, being suitable for the analysis of all types of line patterns. Dacey examined various properties of line patterns in an attempt to find ones which enable patterns of different morphologies to be distinguished one from another. The two obvious candidates, length of line segments and frequency of intersection of a line with other lines, failed in this role. The method Dacey resorts to is to draw a straight line through two arbitrarily located points in the study area and to record the points of intersection of this line with the lines of the pattern. In patterns produced by the Miles model described above we have seen that any line of the pattern can be considered an arbitrary line. Once the points of intersection of the arbitrary line and the lines of the pattern are determined the resulting linear arrangement of points is analyzed using a reciprocal nearest neighbor technique.

Fig. 5.23 shows several kinds of line patterns. Fig. 5.23*a* illustrates a pattern resulting from a process in which pairs of points are located according to Poisson assumptions and straight lines drawn through them. Although the process is not the same as that in the Miles model discussed above it produces

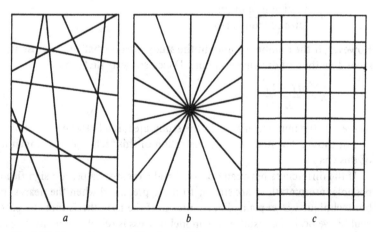

a   *b*   *c*

Fig. 5.23 Line patterns resulting from different generating processes

a pattern which is indistinguishable from that of the Miles model. In particular, both processes produce patterns which could have been created by requiring each unit of the study area to have an equal and independent chance of receiving a line or part of a line. Consequently, statements made about the pattern shown in fig. 5.23*a* are also applicable to patterns produced by the Miles model. The processes underlying the patterns in figs. 5.23*b* and 5.23*c* should be readily apparent.

The reciprocal nearest neighbor test has already been described for points on a two-dimensional plane (see chapter 2, pp. 32–33). In patterns generated by the Miles model (fig. 5.23*a*) the points of intersection of a test line with the pattern can be assumed to be Poisson created so that the proportion of reciprocal nearest neighbors along the test line is $(2/3)^j$ (where j is the order of the nearest neighbor) (Clark, 1956). In patterns of the type shown in fig. 5.23*b* there are fewer reciprocal nearest neighbors than in model-generated patterns, while patterns of the type in fig. 5.23*c* have more reciprocal neighbors than model-generated patterns.

'Random walk' procedure. There might be instances where the number of intersections of an arbitrary line (traverse) and the pattern are few. In such cases Dacey (1967) suggests that the method above be combined with a 'random walk' type procedure. For example, consider the tree-like network in fig. 5.24. Begin by selecting an arbitrary starting point for the traverse, point A. Then select a direction, θ, from a table of random numbers for the interval $0 \leqslant \theta < 2\pi$. Draw a straight line from A, direction θ, and stop when this line reaches the furthest line of the pattern and call this point B. Now repeat the procedure using B as a new starting point and selecting a new value for θ. This can be repeated indefinitely. Traverses which fail to intersect with the pattern can be rejected. Once a requisite sample size has been reached the random

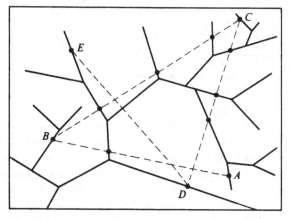

Fig. 5.24 A random walk technique for sampling line patterns

walk can be unfolded to form a continuous straight line traverse. The incidence of intersection points along this traverse can then be examined. See appendix B of this chapter for an illustration of the procedure (pp. 119–120).

Haggett and Chorley (1969, pp. 103–4) have used the technique to analyze patterns formed by the road network in sample areas of south-west England. Haggett calls patterns of the type in fig. 5.23*b* 'radial grids' and those in fig. 5.23*c* 'Manhattan grids'. Most of the road patterns they examined conformed to the expectations of the Miles model.

5.7.   Two-phase mosaics

Pielou (1969, pp. 140–56) has extended the Miles model and used it in the analysis of the areas formed in cell patterns. Although patterns of areas are extensively discussed in chapters 6–8 we shall present her model here since its origins are clearly founded in line-generating processes.

Pielou uses an additional assumption with the basic Miles model. She assumes that each of the cells formed by the lines generated by the model has one of two colors. Each cell is independently assigned a color, black or white, with probabilities b and w, respectively ($b + w = 1$). The pattern so generated is called a two-phase mosaic. A pattern of this kind is shown in fig. 5.25. This

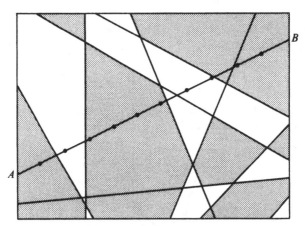

Fig. 5.25 A two-phase mosaic

was created by assigning a color to each of the cells with probabilities $b = w = 0.5$. Suppose the objects to be studied are represented by the black phase and the unoccupied portion of the study area by the white phase. An arbitrary traverse is superimposed on the pattern and the distance of each segment of the line within one color is noted. Pielou gives the expected mean distance of a black segment and a white segment and the respective probability distributions for the length of black and white segments.

Call the length of a black segment, ℓ_b, and a white segment ℓ_w. If λ represents the expected number of network lines cutting across the transect per unit length, then the probability density of ℓ_b is an exponential of the form

$$h(\ell_b) = \lambda_b \exp(-\lambda_b \ell_b) \qquad (5.19)$$

where $\lambda_b = \lambda w$. The expected black segment length is $E(\ell_b) = 1/\lambda_b$. Correspondingly, by substituting w for b, the probability density and expected mean distance of ℓ_w can be obtained.

In practice if it is difficult to find the exact boundaries between b and w an alternative procedure is provided by Pielou. It requires (1) that the end points of a number of traverses be assigned B or W and (2) that the frequency of the combinations of BB, BW, WB, WW be examined for significant differences from the expected combinations.

We should stress that with one transect we are looking at the model in only *one* direction. Ideally, we should demonstrate that the model cannot be rejected in *all* directions. When this can be done Pielou (1965) designates the pattern an *isotropic* one. She also recognizes two other possibilities. A *unidirectional* pattern occurs when the sequences of phases on the traverse fits the expected values in only one direction. If sampling in any direction gives values which fit the expected frequencies but the probabilities vary with direction, the pattern is said to be an *anisotropic* one.

Pielou has extended this approach to cover the analysis of n-phase mosaics, i.e. those instances where the phases represent a number of different species. Readers interested in pursuing this topic should turn to Pielou (1965, 1967, and 1969, pp. 193–200). An application of the technique to the analysis of the spatial locations of groups on a beach is provided by Boots and Merk (1974).

A final tangential point may be in order. MacDougall (1972) has suggested a technique by which any type of mosaic may be generalized so that the original information contained in the mosaic is reduced in the most judicious way. MacDougall's goal was to make a map better perceived and understood by the user. However, his method is useful to those pursuing an analysis of the Pielou type in which there are a large number of phases, many of which are of rare occurrence within the pattern. The approach involves examining the frequency with which one species is contiguous with another and then aggregating those phases which display strong spatial associations. Once the researcher has decided on the number of phases he wishes to reduce the pattern to he has the choice of electing a generalization which either maximizes or minimizes the total perimeter length. Generally, the former emphasizes linear irregular boundaries within the pattern while the latter results in a pattern in which rounded shapes predominate. The latter also eliminates details within the pattern which might otherwise obscure the major trends.

As with the discussion of Dacey's use of the basic line model we must end this discussion of Pielou's work on a cautionary note. The essence of the technique is morphometric analysis. The focus in the examination of the patterns is on their forms rather than on the nature of the processes responsible for their generation.

Appendices to Chapter 5

5.A   An example of the nearest neighbor approach

Consider fig. 5.26. Since the pattern consists of only 9 lines we shall use the second method mentioned on page 113 to derive our sample. To draw

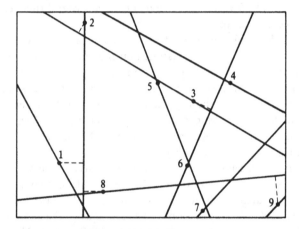

Fig. 5.26 The nearest neighbor technique for a line pattern

the sample we begin by measuring each line. Then we select a point i on each of the lines. This may be done by picking a random number between 0 and 99 from a table of random numbers, multiplying the random number by the length of the line and dividing by 100 to give the distance of i from the end of the line. This means that we must also select the end of the line from which we measure. This can be achieved by designating the digits $(0, 2, 4, 6, 8)$ to represent the north end of the line and the digits $(1, 3, 5, 7, 9)$ to represent the south end of the line. Again we can use a random numbers table to make our selections. The nine sample points selected by this procedure appear in fig. 5.26. We now measure from each of these points the shortest perpendicular distance to a location on any of the eight remaining lines. Summing these perpendicular distances we get 166 units, which gives an average distance, $\bar{r}_{j\,obs}$, of 18.4444 units. To obtain the expected mean distance, $E(r_j)$, we begin by estimating the density of the lines in the pattern. Since our number of lines is small we use expression (5.18) so that the estimated density per unit area, $\hat{\lambda}$, is

$$\hat{\lambda} = \frac{8d}{9}$$

where

$$d = \frac{2512}{(300)(400)} = 0.0209$$

so that $\hat{\lambda} = 0.0186$.
Using (5.16) and (5.17) we get

$$E(r_j) = \frac{0.31831(1)}{0.0186} = 17.1340$$

and

$$v(r_j) = \frac{0.10132}{(0.0186)^2} = 293.5697.$$

Using the expression

$$Z = \frac{\bar{r}_j \text{ obs} - E(r_j)}{\sqrt{(v(r_j)/n)}} \qquad (5.20)$$

where n = number of lines and Z = a normal standard deviate,
we can test if the observed and expected distances to neighboring lines are significantly different. Equation (5.20) yields $Z = 0.2294$ indicating that the difference between the two values is not significant.

5.B   An example of the random walk approach

To illustrate the procedure let us use the line pattern of fig. 5.26. We begin by selecting an arbitrary starting point for the traverse. This may be done by selecting a point at random from within the study area (point A) (see fig. 5.27). If point A is not on a line in the pattern we can begin at A' which is the closest location to A which is on a line in the pattern. We now select a direction, θ, at random. For point A', θ is 271°. Drawing a line orientation 271° from A' we intersect with only one other line of the pattern at point B, which becomes the first hinge on our traverse. For point B we select a new value for θ. The procedure generating the complete traverse of 14 intersections is summarized in fig. 5.27. Whenever, in the course of the procedure, we select a value of θ which produces a traverse segment which does not intersect with the pattern we reject this value of θ and select a new one. The distances between the intersection points of the traverse are given in table 5.5. Analysis of this data shows that there are five pairs of reciprocal points on the traverse

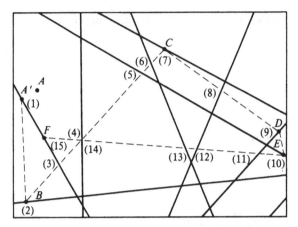

Fig. 5.27 The random walk–reciprocal nearest neighbor technique for a line pattern

which means that the proportion of points that have reciprocal nearest neighbors is 2/3, exactly the proportion we would expect for the Miles model.

Table 5.5. *Interpoint distances on a traverse*

Intersection points	Distance (units)
1–2	144
2–3	75
3–4	53
4–5	126
5–6	17
6–7	31
7–8	102
8–9	101
9–10	31
10–11	32
11–12	92
12–13	18
13–14	149
14–15	57

6

Area patterns: the cell model

As with the point-generated patterns discussed in chapter 5 our concern here is with processes which contain two sets of assumptions. The simplest of these processes are those in which the **P** (point) assumptions are responsible for the location of a set of points and the **A** (area) assumptions involve the creation of a set of areas associated with the points. In this and succeeding chapters we focus our attention on possible **A** assumptions for such processes. Several of the models discussed in these chapters involve **P** assumptions of the Poisson process or some simple modifications of that process. In this chapter we shall confine our attention to an explanation of a basic area model which we call the 'cell' model.

6.1.   Area-generating point processes

In our discussion of line-generating processes in chapter 5 we defined area-generating processes as those in which the **A** assumptions included one which ensured that the point and line components of the pattern formed disjoint sets. We now examine other possible basic assumptions that might be incorporated in area-generating processes. Fig. 6.1 represents the completion of the unfinished branch of fig. 5.1 (p. 87). An assumption may be established which governs the relative time horizon over which the lines (constituting the boundaries of the areas) are generated. The elements of the line set can be created either simultaneously or sequentially. Fig. 6.2 illustrates two patterns of the same form, one of which (that in fig. 6.2a) results from a process in which the lines are created simultaneously, and the other (that in fig. 6.2b) is the product of a process in which the lines are produced sequentially.

Whether the lines are created simultaneously or sequentially a second assumption can determine whether or not the process gives rise to overlapping areas. Fig. 6.3 illustrates two patterns which result from processes which have a common set of **P** assumptions so that the arrangement of the point component is the same in both patterns. However, the **A** assumptions prohibit overlap in the process illustrated in fig. 6.3a but permit it in the process leading to the pattern shown in fig. 6.3b. Clearly, the forms of the patterns are

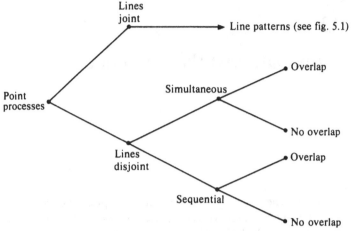

Fig. 6.1 Basic assumptions of area-generating processes

quite different and so too are the respective generating processes. In fig. 6.3*a* the areas are created by assigning all locations in the plane to the nearest member of the point set. In fig. 6.3*b* areas are produced by constructing a circle of fixed radius about each member of the point set. Patterns which result from processes containing assumptions which prohibit the spatial overlapping of the areas will be designated *contiguous* patterns and those which result from processes containing assumptions which permit overlap to occur will be designated *overlap* patterns. Note too that the process modeled in fig.

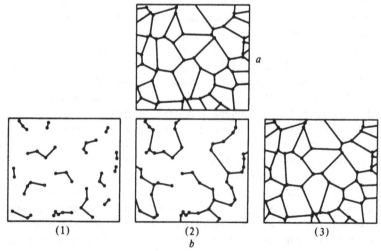

Fig. 6.2 Processes illustrating (*a*) simultaneously generated areas, and (*b*) sequentially generated areas

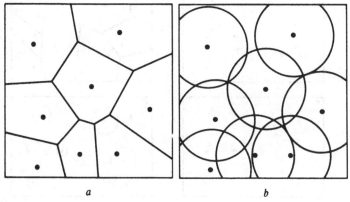

Fig. 6.3 Contrasting assumptions concerning overlapping

6.3*a* leads to a pattern which is *space-exhaustive* (i.e. each location in the plane is assigned to an area). The pattern resulting from the process illustrated in fig. 6.3*b* does not possess this property.

As we stressed in chapter 5, the assumptions outlined in fig. 6.1 are only some of those which could be used in the construction of area-generating processes. Fig. 6.4 shows four patterns evolved under processes incorporating different combinations of the assumptions outlined above. In fig. 6.4*a* all the areas are created simultaneously and overlapping is prohibited. In fig. 6.4*b* areas are again created simultaneously but this time overlapping is permitted. Fig. 6.4*c* and fig. 6.4*d* represent processes in which an assumption of sequential creation of areas replaces one of simultaneous creation. The processes shown in fig. 6.4 are but a few of the number of processes that could be generated. More complex processes can be created if we allow sequential generation of the areas, permitting assumptions to be changed at any time up to the completion of the pattern. An example of such a process is illustrated in fig. 6.5. Here the initial **P** assumptions create a group of points. These are followed by **A** assumptions which lead to the construction of circles about each point thus establishing an initial group of areas. Additional points are created according to a modified group of **P** assumptions which only allow new points outside any of the existing areas. Additional areas are then created by constructing circles around the newly introduced points. The **P** assumptions are further modified to locate additional points outside the existing areas. The same group of **A** assumptions are again used to create further areas. The process terminates when it is no longer possible to locate additional points.

6.2. Properties of area patterns

There are many properties we can examine for patterns which arise as the result of area-generating processes. Some of these are dimensionless such as the number of contiguous neighboring areas (contact number), the number

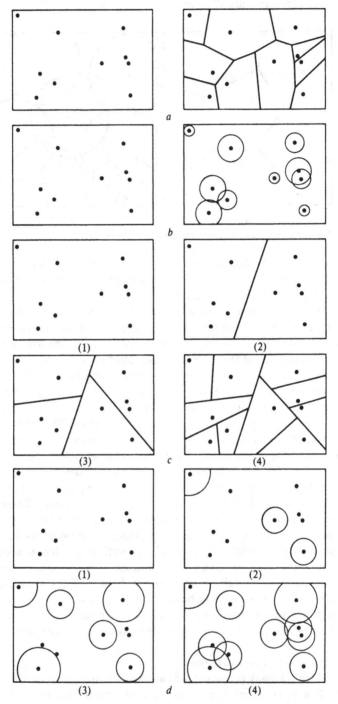

Fig. 6.4 Area patterns resulting from processes involving different combinations of basic assumptions

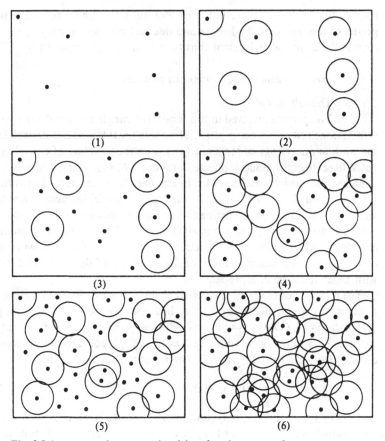

Fig. 6.5 Area-generating process involving changing assumptions

of overlapping areas, and the size of the interior angles if the areas are polygons with straight-line edges.

Other properties are linear (one-dimensional). These usually relate to characteristics of the boundaries of the areas and might include the total length of perimeter or the average edge length. However, linear characteristics need not relate only to the boundary of the area. Examples of properties which do not are the length of longest internal axis ℓ_1, the length of longest axis perpendicular to ℓ_1, and the diameter of the largest circle which can be contained within the figure.

Finally, two-dimensional properties can be noted. These include such things as the physical size of the area, the size of area overlapping with other areas, and the total size of study region covered by more than one area.

Obviously, in any empirical instance, the properties the researcher decides to examine are determined by the nature of the hypothesized process and the

form of the pattern which results. We focus our attention on geometric properties of the type described above and discount non-geometric properties
such as the color, value, internal density, etc. of the areas in question.

6.3.   Processes generating contiguous patterns

6.3.1.   The cell model

The process involved in this model is a simple one involving two **P**
assumptions and one **A** assumption. The name for this model is derived from
the work in mineralogy of Meijering (1953) on the structure of crystal aggregates. In ecology the model is referred to as the 'S-Mosaic' (Pielou, 1969). In an
unnamed form it has also been used in astrophysics to represent instances of
fragmentation (Kiang, 1966). Also unnamed, the model has been presented
with slight modifications to define the 'domain of danger' in the predator-
avoidance behavior of grazing animals (Hamilton, 1971a, b), and to model
cracking in basalt and other materials (Smalley, 1966, 1970). The wide range
of applications is primarily a result of the simplicity of the model and the ease
with which it can be interpreted.

The **P** assumptions are familiar from chapter 2 and consist of locating a set
of points which we label $a_1, a_2, \ldots, a_n$ in a plane by means of a Poisson process of density λ points per unit area. With the point a_i we associate a set of
all locations x in the plane whose distance from a_i is equal to their minimum
distance from the points $\{a_j\}$ $(j \neq i)$. The set of all locations x which satisfy

$$(x - a_i) \leqslant (x - a_j)   (j \neq i) \tag{6.1}$$

produces a set of Thiessen polygons $A_1, A_2, \ldots, A_n$ in which A_i contains all
the locations which are closer to a_i than to any other a_j $(j \neq i)$. It is possible
that a location in the plane, x, is equidistant from a pair of points in which
case x will lie on the boundary of two adjacent polygons. In addition, x may
be equidistant from three points (or in very rare instances, four or more
points) in which case x will form a vertex of three (or four, or more) adjacent
polygons. The resulting polygons (cells) form a contiguous, space-exhaustive
tesselation ($\mathcal{T}$) which is unique for any given distribution of points. A pattern
produced by the operation of the cell model is shown in fig. 6.6. This figure
illustrates the basic characteristics of the cells formed by the process which
are:

(i)   they all have straight line edges;
(ii) all cells are convex (i.e. it is possible to link any pair of points within
      the cell by a straight line without traversing any other cell);
(iii) the minimum number of edges (sides for a cell) is three.

It should be noted that while 'Thiessen polygon' is the name most widely used
in geography these figures are more usually referred to in the mathematical

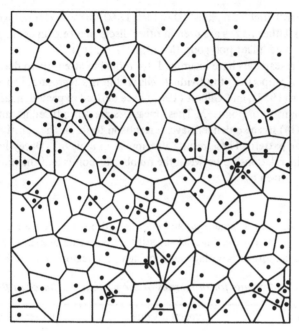

Fig. 6.6 An area pattern generated by the cell model

Table 6.1. *Moments for properties of the cell model*

Property		Expected value
Contact number	$E(N)^a$	6.000
	$E(N)^b$	$8(\pi/6)^{\frac{1}{2}} = 5.7888$
Perimeter	$E(S)^a$	$4\lambda^{-\frac{1}{2}}$
	$E(S)^b$	$5.0133\ \omega^{-\frac{1}{2}}$
Length of any edge	$E(L)^a$	$\frac{2}{3}(\lambda^{-\frac{1}{2}})$
	$E(L)^b$	$(\frac{3}{4}\omega)^{\frac{1}{2}} = 0.86603\ \omega^{-\frac{1}{2}}$
	$E(L^2)^b$	$0.95431\ \omega^{-1}$
Area	$E(A)$	λ^{-1}
	$E(A^2)$	$1.280\ \lambda^{-2}$
Full neighbors	$E(C)$	4.000

λ = expected number of points per unit area.
ω = expected number of vertices per unit area.
[a] After Meijering (1953).
[b] After Dacey (1963*b*).
Sources: Evans (1945); Meijering (1953); Gilbert (1962); Dacey (1963*b*).

literature as either 'Dirichlet regions' (Dirichlet, 1850) or 'Voronoi polygons' (Voronoi, 1908). Other names are used in other disciplines (e.g. in ecology they have been called 'plant polygons' (Mead, 1971, p. 16)).

A number of researchers have examined this model in attempts to derive various geometric properties of individual cells within the aggregate. Evans (1945), Meijering (1953), and Gilbert (1962) have derived expected measures in terms of the density of points, λ. These measures are reproduced in table 6.1. Dacey (1963b) has also derived several results in terms of the density of vertices, ω, in the pattern. These results also appear in table 6.1. Dacey has suggested that the relationship between the number of points and the number of vertices in $\mathcal{T}$ is given by

$$w = c\lambda \qquad (6.2)$$

where w = an empirical estimate of ω and $c = \pi/2$, although evidence from computer simulations of the cell model suggest that the value of $\pi/2 = 1.57$ for c is too small, a more appropriate value being in the region of $c \simeq 1.83$.

Table 6.2. *Moments for properties of the pattern illustrated in fig. 6.6*

Property	Expected value	Observed value
$E(N)^a$	6.000 ⎫	
$E(N)^b$	5.789 ⎬	5.988
$E(S)^a$	9.782 ⎫	
$E(S)^b$	9.015 ⎬	9.696
$E(L)^a$	1.630 ⎫	
$E(L)^b$	1.557 ⎬	1.659
$E(A)$	5.981	5.714
$E(A^2)$	45.787	40.917
$E(C)$	4.000	4.174

$\hat{\lambda}$ = empirical estimate of λ = 0.1672.
w = empirical estimate of ω = 0.3092.
[a] After Meijering (1953).
[b] After Dacey (1963b).

We can use the values in table 6.1 to estimate the properties for the pattern in fig. 6.6. These expected values are given in table 6.2, which also contains the observed values for the pattern in fig. 6.6. It will be seen that there appears to be a good fit between the two sets of values. Note also that we have only measured the properties of the eighty-six polygons which fall entirely within the study area. Any polygons which have part of the study area boundary as one of their edges have been discounted. Regarding the two conflicting values given in table 6.1 for the mean value for the contact number of an individual cell, the existing evidence seems to indicate that the value of 6.0 is the correct one. Indeed, based on the initial work by Graustein (1931), Smith (1952, 1954), Matschinski (1954), and Woldenberg (1970) suggest that in any contiguous, space-exhaustive pattern of cells where three, and only three, edges are incident at any vertex the mean contact number will always be six. In the cell model the incidence of more than three edges at a vertex is a very rare occurrence. For four edges to be incident, four points must be the vertices of

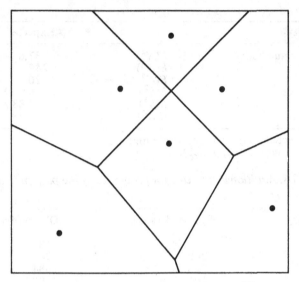

Fig. 6.7 The incidence of four edges at a vertex in $\mathcal{T}$

a square or a rectangle (see fig. 6.7). Thus, the value of six for the mean contact number in this instance would appear to be the appropriate one. Dacey (1963), Haggett (1965), Pedersen (1967) and Boots (1970) are among those who have studied mean contact number values for patterns occurring in geography. For some socio-economic networks Boots (1970) reports values ranging from 6.00 (for the non-coastal, non-border states of the U.S.A. and ecclesiastical parishes in Wiltshire, England) to 5.35 (for counties in New South Wales). Similar results have been reported for natural phenomena (e.g. Lewis, 1946).

It will be noted that the measures listed in table 6.1 are almost all limited to mean values. This is because of the difficulty of deriving higher-order moments analytically. This has led several researchers to turn to computer simulation approaches. Using such an approach Crain (1972) has made estimates of several higher order moments from information obtained for 11 000 polygons generated by a Monte Carlo simulation procedure. Crain's estimates are given in table 6.3. These estimates may be compared with the

values obtained for the polygons of the pattern in fig. 6.6 which are
reported in table 6.4. In addition, Crain has obtained an estimate of the
probability distribution of the number of k-sided polygons generated
by the cell model (see table 6.5). Table 6.6 compares the frequencies
of k-sided polygons for the pattern in fig. 6.6 with Crain's estimates using

Table 6.3. *Estimates of higher order moments for properties of the
cell model*

Property		Expected value
Contact number	$E(N^2)$	37.8
	$E(N^3)$	249
	$E(N^4)$	1720
Perimeter	$E(S^2)$	16.96 λ^{-1}
	$E(S^3)$	74.88 $\lambda^{-\frac{2}{3}}$
Area	$E(A^3)$	1.78 λ^{-3}

λ = expected number of points per unit area.
Source: Crain (1972, p. 220).

Table 6.4. *Additional moments for properties of the pattern illustrated in fig. 6.6*

Property	Expected values	Observed values
$E(N^2)$	37.8	37.6
$E(N^3)$	249	246
$E(N^4)$	1720	1684
$E(S^2)$	101	98.8
$E(S^3)$	1095.2	1050.5
$E(A^3)$	380.8	355.0

$\hat{\lambda}$ = empirical estimate of λ = 0.1672.

Table 6.5. *Crain estimates of the frequency distribution of k-sided
polygons for the cell model*

Number of polygon sides (k)	Estimated proportion
3	0.011
4	0.111
5	0.259
6	0.288
7	0.206
8	0.087
9	0.029
$\geqslant 10$	0.009
	1.000

Source: Crain (1972, p. 220).

Table 6.6. *Crain estimates and the observed frequency of contact numbers for the pattern illustrated in fig. 6.6*

Number of polygon sides (k)	Expected frequency	Observed frequency
3	0.946 ⎫	0
4	9.546 ⎭	11
5	22.274	21
6	24.768	28
7	17.716	14
8	7.482 ⎫	8
9	2.494 ⎬	4
⩾10	0.774 ⎭	0
(N)	86	86

$X^2 = 1.444$ $df = 4$ $\chi^2_{0.20} = 5.989$

Table 6.7. *Kiang estimates of cell area and observed areas for the pattern illustrated in fig. 6.6*

Standardized area of cell ($A\lambda$)	Observed frequency	Observed proportion	Observed cumulative proportion	Expected proportion[†]	Expected cumulative proportion
0–0.344	5	0.0581	0.0581	0.0510	0.0510
0.345–0.744	23	0.2674	0.3255	0.2962	0.3472
0.745–1.144	33	0.3837	0.7092	0.3229	0.6701*
1.145–1.544	16	0.1861	0.8953	0.1933	0.8634
1.545–1.944	5	0.0581	0.9534	0.0872	0.9506
1.945–2.344	2	0.0233	0.9767	0.0330	0.9836
2.345–2.744	2	0.0233	1.0000	0.0114	0.9950
>2.744	0	0.0000	1.0000	0.0050	1.0000
(N)	86				

*$D_{obs} = 0.0391$ $D_{0.20} = 0.1154$

$\hat{\lambda}$ = empirical estimate of λ = 0.1672.
†Source: Kiang (1966).

a X^2 test. The result indicates a good fit between the two sets of data: observed $X^2 = 1.444$, while χ^2 at the 0.20 level of significance is 5.989 with $4\,df$.

Kiang (1966), also using a computer simulation approach, has speculated that the distribution of sizes of cells in $\mathcal{T}$ is given by a gamma distribution (see pp. 50–1 for a discussion of the gamma distribution), so that

$$g(A; 4, 4\lambda) = \frac{4}{\Gamma(4)} (4A\lambda)^{4-1} e^{-4A\lambda} \tag{6.3}$$

This gives a value of $E(A^2) = 1.250\,\lambda^{-2}$ which is close to the value derived analytically by Gilbert (1962) (see table 6.1). In order to make this information more readily usable by the reader, table 6.7 gives the distribution of a measure of standardized area, $A\lambda$ (where A = area of cell). We may use this information to examine the distribution of the sizes of the cells in fig. 6.6. Table 6.7 shows that a K–S test reveals a close correspondence between the values for the cells in fig. 6.6 and Kiang's expectations.

We have already indicated that the cell model has been used in disciplines ranging from astrophysics to biology and noted that this wide ranging utility is due to the basic simplicity of the model. In particular, the model can be used to describe two distinct processes. The simplest is an 'assignment' process in which portions of the plane are allocated to points. The assignment procedure in the cell model is one in which each location in the plane is allocated to the point closest to it. Bogue (1949) used Dirichlet regions defined around a set of 67 metropolitan centers within the United States as proxy for real market areas arguing that, as a rule-of-thumb, a metropolitan center could be expected to dominate all locations that were closer to it than to any other metropolitan center. Although Bogue's approach used Thiessen polygons it did not represent an application of the cell model since the locations of the metropolitan centers were already determined and clearly not the result of a Poisson process. Several archaeologists have also used variations of the model in similar ways in attempts to reconstruct past territorial patterns (e.g. Clarke, 1972; Hammond, 1972; Hodder, 1972).

The second basic process represented by the model is a simple 'growth' process. For crystal aggregates Meijering (1953) gives the model an interpretation by considering the points as nuclei for the growth of crystals. His concern is primarily with three-dimensional forms although no difficulty is encountered in interpreting the model if the analysis is confined to two dimensions. In such instances the points (a_i) can be viewed as the original nuclei, centers, focal points, nodes, energy input points, seeds, etc. of the resulting cells A_i. If this is so the following **P** and **A** assumptions are implicit for the growth process:

(i)   all nuclei appear simultaneously and remain fixed in terms of their location in the plane throughout the growth period;

(ii) all nuclei are of equal value;

(iii) growth occurs in all directions at the same rate;

(iv) growth ceases when each cell A_i comes into contact with all neighboring cells A_j $(j \neq i)$.

The cell model has been used by geographers in several instances. Dacey (1963) used the model essentially in an assignment role examining contact number properties for patterns of market areas. Boots (1972, 1973) gave the model a 'growth' interpretation and used it together with the Johnson-Mehl model (see pp. 145–51) in an investigation of bus service center hinterlands in England and Wales. The hinterlands used were those delimited by Green (1955), and two sample study areas contrasting in physical and socio-economic

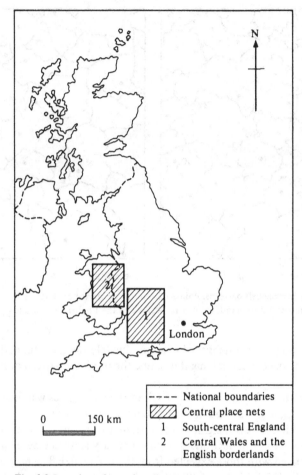

Fig. 6.8 Location of sample cell patterns

conditions were chosen. These areas were south-central England, and central Wales and the English borderlands (see fig. 6.8). Both patterns are established in predominantly rural areas but the English area has a much higher population density than the Welsh area. This reflects contrasts in the underlying physical structure of the two areas. The English area is almost entirely part of the fertile Midland Plain while the Welsh area for the most part covers a predominantly barren upland area. The two sample networks are illustrated in fig. 6.9. In Great Britain bus services have their origins in the 1920's and 1930's when they first challenged the hegemony of the railways. The development of these services was not coordinated by any central government agency and the process was a highly competitive one. Bus services became established at existing central places. If the centers were located following Poisson assump-

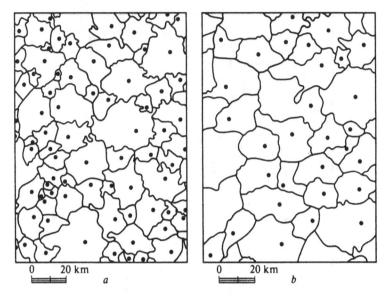

Fig. 6.9 Sample cell patterns. Points represent bus service centers for:
(a) south-central England; (b) central Wales and the English borderlands

tions and bus services were established at approximately the same time, the growth version of the cell model might describe the resultant pattern of hinterlands.

First, the location of central places in each of the study areas was examined using a nearest neighbor test, and in both instances the centers were found to have locations which were not significantly different from ones generated by a Poisson process. For both networks selected summary statistics were computed and compared with those expected from the cell model. The results are given in table 6.8. This table shows that the cell model is only rejected for one property (edge length) for one network (England).

Table 6.8. *A test of the cell model using data on bus service hinterlands taken from fig. 6.9*

Property	Observed values		Expected values		Test for significant difference	
	England	Wales	England	Wales	England	Wales
$E(N)$	5.711	5.600	6.000	6.000	$t = 0.915$	$t = 1.347$
$E(S)$	43.622	47.350	40.000	56.569	$t = 1.008$	$t = 1.896$
$E(L)$	7.631	8.500	6.667	9.428	$t = 2.551*$	$t = 1.485$
$E(A)$	95.731	125.950	100.000	200.000	$t = 0.010$	$t = 0.027$
$E(A^2)$	16727	21268	12800	51200	–	–

* Difference significant at the 95 per cent level.

However, based on the assumptions associated with a growth interpretation of the model (see above) it is somewhat surprising that we get such a good fit. For many empirical instances the artificiality of the growth assumptions make the model of limited use as an explanatory tool. However, the model provides a useful standard by which to evaluate real world patterns in a manner similar to the role sometimes played by the Poisson distribution for point patterns. It can serve to indicate in which ways an observed pattern differs from that expected under the constraints of the cell model, providing insights into which of the non-geometric characteristics of the pattern have been responsible for producing the observed changes in the geometric characteristics.

Thus, given the limitations imposed by the assumptions of the model, it might prove fruitful to examine modifications that can be made to the basic cell model in order to make it appropriate in certain specific instances. Since the cell model involves two sets of assumptions there are two major kinds of modifications which may be undertaken. These concern the **P** assumptions, responsible for the location of points, and the **A** assumptions, responsible for the development of the areas.

In the basic model described above the points are located according to a Poisson process. It is possible to locate the points in the plane according to other probability models. In chapter 3 we encountered a range of alternative distributions which are useful in geography. In the appropriate circumstances we might consider it desirable to replace the assumptions of the Poisson process with those of other processes. In addition, we might develop our own rules for constructing the areas. For instance, we could account for upper and lower threshold constraints by specifying maximum and minimum distances between neighboring points. Such distances may be kept constant or made variable. The latter could well be desirable if we choose to differentiate the points in terms of size or income, say, as we might wish to do if we were using the model to describe central place hinterlands established around centers of differing functional size. If relocation of points occurs in response to threshold demands, putting upper and lower limits on the distance between points will have the effect of bringing about a trend toward the regular spacing of points. If this is so, and the other conditions of the basic cell model are retained, the effect will be to reduce the variance associated with all the measures of the geometric characteristics of the model. Perhaps the most systematic approach to changes in the location of points would be to examine the characteristics of patterns developed according to the procedures of the cell model around a set of points which are initially regularly spaced but in which the locations of the points become successively disturbed until they are indistinguishable from locations chosen following Poisson assumptions.

Second, we may consider modifications to the **A** assumptions of the process. In the basic cell model the partitioning is achieved by creating Thiessen polygons around the points. In this way all locations in the plane are assigned to their nearest point. The resulting cellular net pattern is unique. However, it might be desired that locations are not assigned to their nearest point. This might be so if points are differentiated so that the more important points might exert additional pull thus extending their boundaries at the expense of less important neighboring points. Cox and Agnew (1974) used such a model to examine the evolution of the thirty-one counties of Ireland. Another reason for a revised assignment process might be the existence of barriers in the plane. However, in some instances a unique tesselation of polygons cannot be achieved unless additional assumptions are made.

The most rewarding approach might be to retain the cell model as a basic building block and to modify it for specific empirical circumstances. This is the approach taken by Smalley (1970). We shall use his work on crack patterns in basalt as an example (Smalley, 1966).

Cracks in basalt result from contraction of lava flows on cooling. This contraction causes tensile stresses in the rock mass which in turn causes cracks. In material of uniform temperature a regular arrangement of stress centers occurs and the crack network takes on an hexagonal form. However, if uniformity of temperature is not present, as is usually the case, the cracks form independently of each other around a set of stress centers Poisson-like in their location. Smalley uses a simple modification of the cell model to describe such occurrences. Points are selected at random in the plane subject to the constraint that points which would be closer to already selected points than a specified distance are rejected. The points are considered stress centers (points around which tensile stresses develop) and so the specified critical distance is equal to twice the radius of the stress circles. Points are sequentially generated until no more can be located without violating the interpoint distance constraint. This is equivalent to ensuring a minimum threshold size for each cell. In this context the cell model is equivalent to the sequential packing of circles in a plane, a problem which has been studied analytically for some time in geometric probability without much success (Kendall and Moran, 1963).

Once the points have been located Smalley constructs the areas in the same way as in the basic cell model, i.e. he constructs a set of Thiessen polygons. An example of one of the patterns produced in this way is illustrated in fig. 6.10. Smalley only examined the contact number properties of the trial networks he developed. As expected the effect of the constraint is to reduce variability in values. Summaries of his values are given in column one of table 6.9. Smalley found a preponderance of pentagons and hexagons in the model generated patterns which was in agreement with values observed in real-world basalt flows reported by Beard (1959). However, sufficient discrepancies exist between model results and those of real world basalt patterns that a X^2 test leads to the rejection (at the 0.20 significance level) of the Smalley model for each of the four empirical patterns (see table 6.9). Smalley explains these differences by pointing to the existence of very short edges in the model pat-

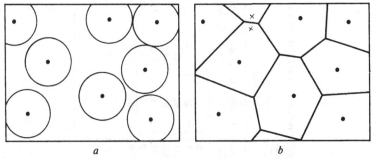

Fig. 6.10 Pattern produced by a modification of the cell model. (*a*) Location of stress centers. (*b*) Creation of areas (short side is indicated by crosses)

terns (see fig. 6.10). The counterparts of such edges in real-world lava flows may well have been overlooked in the compilations of field counts. Smalley thus suggests ignoring the short edges occurring in his model patterns, and recalculating the contact numbers (these revised values are given in column two of table 6.9). The new values generally offer a better fit to observed values, but again a X^2 test leads to the rejection (at the 0.20 significance level) of the Smalley model.

The results of Smalley, together with those already reported for comparisons with the basic cell model, suggest that only rarely will the conditions of the cell model be encountered in reality and that its major role is that of a standard. Perhaps the major shortcoming of the model is the absence of a time dimension. Although we can view the model in a growth context the fact that all the points are located simultaneously and growth occurs at a constant rate in all directions is equivalent to a situation in which all the cells are created instantaneously.

6.3.2. Delaunay triangles

Whenever we create the tesselation of Thiessen polygons $\mathscr{T}$, for a set of points we also automatically generate another space-exhaustive, contiguous tesselation. This 'dual' tesselation is the set of Delaunay triangles, $\mathscr{D}$. Properties of this tesselation can be used in some instances in point pattern analysis as an alternative to nearest neighbor and quadrat techniques to test the hypothesis that an observed pattern could have been the result of a Poisson process (Boots, 1974, 1975*a*). Whether or not the properties of $\mathscr{D}$ can be determined for generative processes other than the Poisson has not yet been determined.

If we examine the vertices of $\mathscr{T}$ (see fig. 6.6) we see that three edges are incident at each vertex. We have already noted that the incidence of more than three edges at a vertex is a rare occurrence in patterns resulting from a Poisson process. Indeed, for non-uniform point patterns (especially the ones discussed here) the occurrence of four edges at a vertex has such a low prob-

Table 6.9. *Expected proportions of contact numbers using original and modified Smalley model: tests on basalt cracks at four sites*

Contact number	Expected (1)	Proportions (2)	Lewiston (Idaho)			Giant's Causeway (Northern Ireland)			Devil's Postpile (E. California)			Dunsmuir (California)		
			Observed	Expected (1)	Expected (2)	Observed	Expected (1)	Expected (2)	Observed	Expected (1)	Expected (2)	Observed	Expected (1)	Expected (2)
3	0.0000	0.0000	0	0 ⎫	0 ⎫	0	0 ⎫	0 ⎫	2	0 ⎫	0 ⎫	1	0 ⎫	0 ⎫
4	0.0392	0.0980	5	2.6 ⎬	6.6 ⎬	18	15.7 ⎬	39.2 ⎬	38	15.7 ⎬	39.2 ⎬	29	7.8 ⎬	19.6 ⎬
5	0.2353	0.2549	30	15.8	17.1	140	94.1	102.0	150	94.1	102.0	92	47.1	51.0
6	0.4706	0.5098	28	31.5	34.2	204	188.3	203.9	178	188.3	203.9	67	94.1	102.0
7	0.1961	0.1373	4	13.1 ⎫	9.1 ⎫	37	78.4 ⎫	54.9 ⎫	32	78.4 ⎫	54.9 ⎫	9	39.2 ⎫	27.4 ⎫
8	0.0588	0.0000	0	4.0 ⎭	0 ⎭	1	23.5 ⎭	0 ⎭	0	23.5 ⎭	0 ⎭	2	11.8 ⎭	0 ⎭
(N)			67	67	67	400	400	400	400	400	400	200	200	200
X^2				25.4	14.2		67.4	30.9		122.5	35.5		144.8	60.4
df				1	2		3	2		3	2		3	2
$\chi^2_{0.20}$				1.6	3.2		4.6	3.2		4.6	3.2		4.6	3.2

(1) Original Smalley model.
(2) Modified Smalley model.
Calculated using data from Beard (1959) and Smalley (1966).

ability that it may be discounted with impunity. If we stipulate that three, and only three, edges meet at a vertex in $\mathcal{T}$, then each vertex is equidistant from three of the points generated by the Poisson process and is common to three Thiessen polygons.

Delaunay triangles can be constructed by joining those pairs of points a_i, a_j whose Thiessen polygons have an edge in common. Clearly, these points are those which are equidistant from the vertex produced by the common edges. The result is a second non-overlapping, space-exhaustive, unique tesselation of triangles, $\mathcal{D}$. $\mathcal{D}$ can thus be considered the geometric dual of $\mathcal{T}$. $\mathcal{D}$ corresponding to the $\mathcal{T}$ in fig. 6.6 is illustrated in fig. 6.11.

Clearly, the two tesselations are closely related. Miles (1970, p. 108) has shown that if there are n_i polygons in the interior of $\mathcal{T}$ and n_b boundary polygons there are

$$2n_i + n_b - 2 \tag{6.4}$$

triangles in $\mathcal{D}$. In addition, the number of edges incident at any vertex in $\mathcal{D}$ will equal the number of sides of the polygon in $\mathcal{T}$ in which the vertex is a

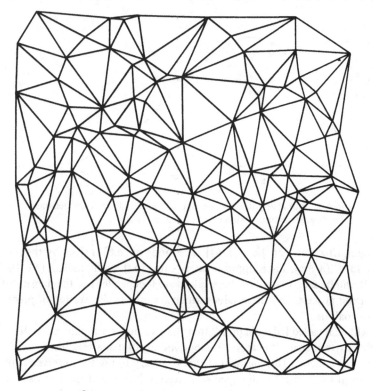

Fig. 6.11 $\mathcal{D}$ for the pattern illustrated in fig. 6.6

node. Finally (using the superscript prime to indicate the complementary angle in π), if the interior angles (in radians) of a polygon at a vertex in $\mathcal{T}$ are α', β', γ', then α, β, γ are the corresponding angles of $\mathcal{D}$. We make use of this property in our analysis.

Miles (1970, pp. 111–13) has done considerable work on the probability distribution of random variables describing characteristics of $\mathcal{D}$. In particular, he has derived the probability density function (PDF) of a pair of angles selected at random from an arbitrary triangle in $\mathcal{D}$. This is

$$f(\alpha, \beta) = \frac{8}{3\pi} \sin \alpha \sin \beta \sin (\alpha + \beta) \quad (\alpha > 0, \beta > 0, \alpha + \beta < \pi) \qquad (6.5)$$

The mode of (6.5) is $\alpha = \beta = \pi/3$ indicating that the most likely triangle in $\mathcal{D}$ is an equilateral one. By integrating (6.5) over all values of β, Miles obtains

$$f(\alpha) = 4[(\pi - \alpha) \cos \alpha + \sin \alpha] \frac{\sin \alpha}{3\pi} \quad (0 < \alpha < \pi) \qquad (6.6)$$

as the PDF for an angle selected at random from $\mathcal{D}$. This distribution is shown in fig. 6.12. The mean of α is, of course, $\pi/3$ and the second moment about

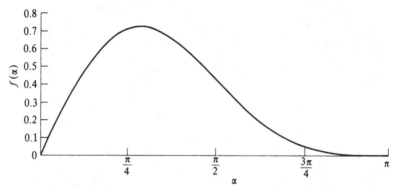

Fig. 6.12 Probability density function of a randomly selected Delaunay triangle angle. Source: Miles (1970, fig. 3, p. 112)

the origin, $E(\alpha^2)$, is $[(2\pi^2)/9 - (5/6)]$. The mode of (6.6) is given by the root of $\tan 2\alpha = 2(\alpha - \pi)$, and is approximately equal to 0.89489 radians ($51°16'$ $24''$). It will be seen that all of these values are independent of the density of the Poisson process. It is these values which are used in this version of point pattern analysis.

It will be recalled from chapter 2 that the two major sampling procedures of point pattern analysis nearest neighbor and quadrat analysis, were both 'density dependent'. This means that it is necessary to use the observed density of points in the derivation of expected pattern values which are then used to test observed patterns. Density dependent values have also been derived by

Table 6.10. *Density dependent values of* $\mathscr{D}$

Property		Expected value
Perimeter	$E(S)$	$\frac{32}{3\pi}\lambda^{-\frac{1}{2}} = 3.395\,\lambda^{-\frac{1}{2}}$
	$E(S^2)$	$\frac{125}{3\pi}\lambda^{-1} = 13.263\,\lambda^{-1}$
Area	$E(A)$	$\frac{1}{2}\lambda^{-1} = 0.500\,\lambda^{-1}$
	$E(A^2)$	$\frac{35}{8\pi^2}\lambda^{-2} = 0.443\,\lambda^{-2}$

λ = expected number of points per unit area.
Source: Miles (1970).

Miles (1970) for properties of $\mathscr{D}$ generated by a Poisson process. These values are given in table 6.10 together with estimates of second order moments.

We may use the 'density-free' properties of $\mathscr{D}$ to test if a given point pattern is significantly different from one generated by a Poisson process in the following way. Given the pattern of points, we generate $\mathscr{T}$ associated with the pattern. Because of the relationship between the angles of the triangles in $\mathscr{D}$ and the interior angles of the polygons in $\mathscr{T}$ it is not necessary to construct $\mathscr{D}$. We simply sample in an equally likely way the set of observed values of the complements in π of the interior angles of the polygons in $\mathscr{T}$. These values are cumulated and the frequency distribution compared with the expected one derived from the indefinite integral of (6.6). The probability that an arbitrarily selected angle assumes a value less than or equal to x is

$$F(\alpha) = \int_0^x f(\alpha)\,\mathrm{d}\alpha$$

$$= (1/3)\left[2\sin^2\alpha + \frac{1}{\pi}\left(\alpha\cos 2\alpha - \frac{3\sin 2\alpha}{2} + 2\alpha\right)\right]_0^x \qquad (6.7)$$

Goodness-of-fit can be evaluated using a K–S test. This procedure provides a test, at whatever confidence level is desired, of the hypothesis that the original point pattern conforms to one generated by a Poisson process. The entire procedure has been automated by modifying and extending the program 'THIESEN' (Rhynsburger, 1973) and is illustrated for the point pattern shown in fig. 6.13. This pattern is a portion of a larger pattern of points generated using coordinates selected from a random numbers table. Table 6.11 shows that, as we might expect, there is a good fit between the observed and expected values of $F(\alpha)$.

However, if we had rejected the Poisson model, nothing further could have been said. In the nearest neighbor and quadrat approaches, even with their shortcomings, rejection of the model sometimes allows one to speculate about the nature of alternative generating processes. Research has not yet advanced

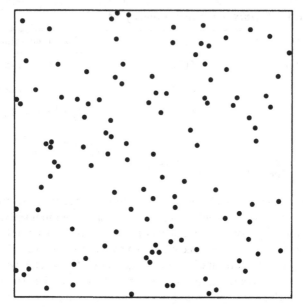

Fig. 6.13 Point pattern analyzed by the use of $\mathscr{D}$

to the position at which we can see if the use of Delaunay triangles will also allow us to make such inferences when the null hypothesis of a Poisson generative process is rejected. Clearly, this represents an interesting research area. Boots (1975a) presents some preliminary findings.

6.3.3.   Information theory and the cell model

The information content of a pattern may also be studied using the cell model. Instead of using probabilities to represent the frequency of points

Table 6.11. *Frequency distribution of α for the sample point pattern of fig. 6.13*

(α)	Observed cumulative frequency	$F(\alpha)$ Observed	Expected
$<\pi/8$	18	0.1035	0.0979
$<\pi/4$	69	0.3966	0.3409*
$<3\pi/8$	105	0.6036	0.6181
$<\pi/2$	139	0.7989	0.8333
$<5\pi/8$	165	0.9484	0.9510
$<3\pi/4$	172	0.9886	0.9925
$<7\pi/8$	174	1.0000	0.9998
$<\pi$	174	1.0000	1.0000

* $D_{obs} = 0.0557$ $D_{0.20} = 0.0811$

in quadrats as was done in chapter 4, here we use probabilities based on the size of Thiessen polygons. Chapman (1970) suggests that we take the Thiessen polygon surrounding a point as a proportion of the total area. The set of these proportions (p_i's) is evaluated as before (see pp. 81–5 of chapter 4).

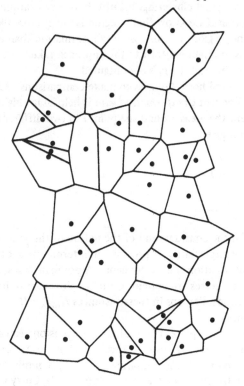

Fig. 6.14 An area pattern generated by the cell model for use in information theory analysis

As an example let us consider the diagram shown in fig. 6.14. This was created by using the assumptions of the cell model except that areas distorted by the border effect are not shown. Consequently, the forty-five polygons (subareas) represent the entire study area. After finding the proportion of the total area for each subarea we use (4.9) and obtain the value for H of 3.688. This is then compared to the maximum of H, H_{max}. Maximum uncertainty results when all polygons are of equal size. By taking $1/n$ as our measure for each p_i we get H_{max} to be 3.807. Lenz (1977) has found that the expected difference between a Poisson process pattern and H_{max} is approximately 0.124 no matter what value n takes. In our example the difference is 0.119. As a result we can again see that a Poisson generated pattern creates a value close to what we have been calling maximum uncertainty.

In a highly clustered point pattern the great number of small polygons created around the points contrasts with the extremely large polygons in the relatively pointless areas. The variety of high and low p_i values yields a low H. Consequently, we can evaluate the degree of clustering by the measure $H_{max} - H$. High values imply great clustering. Intuitively one can rationalize this measure by assuming that if we know that a region is much more likely to receive a point than another region, we possess more information than otherwise. This interpretation is in direct contrast to the approach taken in chapter 4. The difference is due to the way the p_i's are obtained.

The method of analysis used here allows us to make comparisons of H for several time periods. The concept of *information gain* is helpful in this regard. Suppose x_i and y_i represent the areas around the point i in two different time periods. Information gain is given as:

$$H_G = y_i \left(\ln \frac{1}{x_i} - \ln \frac{1}{y_i} \right)$$

$$= y_i \ln \frac{y_i}{x_i} \tag{6.8}$$

The measure H_G allows for the consideration of the nature of the process as it is 'in progress' and, more importantly, it provides an interpretation of pattern change in terms of information theory concepts. A problem arises, however, when birth or death processes are present. Certain points cannot be traced through all time periods. Perhaps in these instances $H_{max} - H$ can be used as a measure for comparison.

Theil (1967), an economist, explores many of the measures one can adapt for use in this context. Chapman (1970) shows several techniques for eliciting further information from point patterns. Empirical work in geography on information theory can be found in Berry and Schwind (1969), Curry (1972), Semple (1973), Garrison and Paulson (1973) and Chapman (1973). Batty (1974) concerns himself with the problem of developing regions via measures of information and Medvedkov (1970) assesses the concept of entropy for use in geography. The term 'entropy' runs through all of these studies, but has been carefully avoided in this book for fear that its several definitions would confuse rather than enlighten in this introductory treatment.

7
Area patterns: the Johnson–Mehl model

In this chapter we continue our discussion of processes which produce contiguous patterns of areas with an examination of the relatively complex Johnson–Mehl (J–M) model. This model was originally developed in relation to the growth of crystal aggregates (Johnson and Mehl, 1939) and has also been used in the examination of the growth of surface films on metals (Evans, 1945).

In the model, the assumption of simultaneous appearance of points (nuclei) made in the cell model is replaced by one in which, starting at an initial time $t = 0$, nuclei appear at a constant rate of α nuclei per time unit per unit area. The rate of nucleation, α, represents the change in the number of nuclei (dN) per time period (dt), i.e.

$$\alpha = \frac{dN}{dt} \tag{7.1}$$

Each nucleus is of the form (A_i, t_i) where A_i is its position in the plane and t_i $(t_i \geqslant 0)$ is its arrival time. Throughout the process each nucleus is considered to be located according to a Poisson process in the plane. Once born, each nucleus grows into a cell at a constant rate of radial growth, v. Both α and v are assumed to be the same for all cells and to remain constant throughout the process. The development of a contiguous cell pattern under the assumptions of this model is illustrated in fig. 7.1.

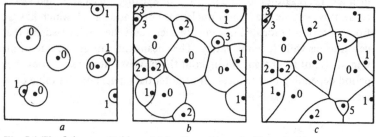

Fig. 7.1 The Johnson–Mehl model for times (a) $t = 1$, (b) $t = 3$, (c) $t = 7$

An unimpeded cell will grow equally in all directions as a circle from its origin at the parent nucleus (A_i). The resulting rate of cell face expansion of the unimpeded cell is directly proportional to the square of the radial growth. At time t, the size of the cell, s, will equal $\pi v^2 t^2$ and so

$$\frac{ds}{dt} = 2\pi v^2 t \tag{7.2}$$

where $\frac{ds}{dt}$ = rate of cell growth. However, if we again look at fig. 7.1 we will see that as the cells grow there is a mutual impedance of growth. The model assumes that cell growth ceases wherever two or more cells impinge. Subsequent to impingement, the rate of change of the size of the cell will be reduced to a proportion of the original unimpeded rate. On average this proportion is equal to the portion of the untransformed plane (i.e. unoccupied portion of the plane) at time t. Thus,

$$\frac{ds}{dt} = u(t)2\pi v^2 t \tag{7.3}$$

where $u(t)$ = proportion of plane untransformed at time t. The value of $u(t)$ can be obtained by multiplying the rate of growth of one cell by the number of points nucleated at the same time. This gives the rate of growth of all points nucleated at this particular time. If we then sum such expressions for all times of nucleation we obtain the rate of transformation. Integrating this expression gives the equation for the fraction of the plane transformed as a function of time (Johnson and Mehl, 1939, p. 420). Thus

$$u(t) = \exp\left(-\pi\alpha v^2 t^3/3\right) \tag{7.4}$$

A plot of (7.4) for increasing values of t reveals an 'S-shaped' curve indicating that soon after the process begins the transformation is rapid and later levels off. Note also that the amount of untransformed plane is affected more by the rate of growth, v, since it is squared, than by the nucleation rate, α.

Since the model assumes that the nuclei are located in a Poisson process manner (i.e. regardless of the position of existing nuclei and their associated cells), it is possible that a new nucleus is born inside a cell which has grown from a previously located nucleus. In such cases the model assumes that the new nucleus will die immediately. This means that the actual rate of nucleation, α', is not a constant throughout the process but instead varies each time period depending on the location of potential nuclei and the value of $u(t)$. In fact,

$$\alpha' = u(t)\frac{dN}{dt} \tag{7.5}$$

At the completion of the colonization process the number of nuclei per unit area of the plane is given by (Gilbert, 1962, p. 962) the expression

$$(\alpha/v)^{D/(D+1)}\{(D+1)/S(D)\}^{1/(D+1)}\Gamma([D+2]/[D+1]) \qquad (7.6)$$

where D = dimension of the pattern and

$$S(D) = \pi^{D/2}/\Gamma(D/2+1)$$

For two-dimensional patterns of the kind we are interested in, (7.6) reduces to

$$0.8794(\alpha/v)^{2/3} \qquad (7.7)$$

Each successfully established nucleus, A_i, eventually produces a cell, C_i, containing all locations, L_k, such that the cell growing from A_i is the first one to reach L_k. This can be expressed in terms of the time taken to reach L_k, as follows

$$t_i + (r_i/v) < t_j + (r_j/v) \qquad (j \neq i)$$

where r_i and r_j are the distances from L_k to A_i and A_j, respectively.

If we again turn to fig. 7.1 and examine the cells produced we see that in terms of their morphology they are different from those cells produced by the cell model (see fig. 6.6).

In particular, their major characteristics are:

   (i)   curvilinear and straight-line edges;
   (ii) cells not necessarily convex;
   (iii) the minimum number of edges for any cell is two which implies;
   (iv) the common edge between two adjacent cells can be discontinuous.

Because of its potential value to geographers it is unfortunate that it is even more difficult to derive analytically properties of patterns produced by the J–M model. Analytically derived results are currently limited to those given in table 7.1. However, we can use these results in pattern analysis. Fig. 7.2 is a pattern generated using the assumptions of the J–M model. In this case eight points are located in a Poisson process manner within a 10 x 10 unit grid during each time period as potential sites for nuclei (thus, $\alpha = 0.08$). Those nuclei which are located in unoccupied portions of the plane generate cells which grow at a constant linear rate of 0.2 units per time interval in all directions in which their growth is unimpeded (thus, $v = 0.2$). Table 7.2 gives the expected and observed values for properties of the cells in fig. 7.2. As usual those cells which contact the boundary of the study area are omitted from the analysis.

One problem encountered in the use of model properties in pattern analysis is that the expected values are expressed in terms of α (the nuclea-

Table 7.1. *Moments for properties of the Johnson–Mehl model*

Property		Expected value
Number of sides	$E(N)$	$<6.000^*$
Perimeter	$E(S)$	$3.734\,[1.137\,(v/\alpha)^{\frac{2}{3}}]^{\frac{1}{2}}$
Edge length	$E(L)$	$<0.627\,[1.137\,(v/\alpha)^{\frac{2}{3}}]^{\frac{1}{2}}{}^*$
Area	$E(A)$	$1.137\,(v/\alpha)^{\frac{2}{3}}$
Full neighbors	$E(C)$	4.000

v = normal velocity of cell growth.
α = density of nuclei arrivals.

* The inequality is necessary because it is possible for two cells to
have more than one common boundary if their 'mathematical boundary' is intercepted by smaller cells.
Source: Evans (1945), Meijering (1953).

tion rate) and v (the velocity of lateral growth) (see table 7.1). This means
that if we are to use such values in analyzing empirical patterns we must be
able to determine values for α and v (as we did above). If we lack sufficient
information on the evolution of the pattern in question this may prove
impossible. Ways of avoiding this situation are to use the result in expression
(7.7), or to use characteristics of the model which are independent of α and
v. In the latter instance, we can use the dimensionless coefficient of vari-

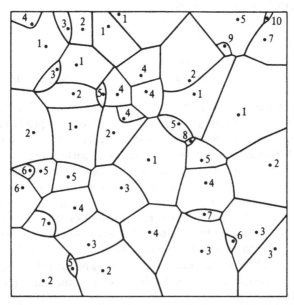

Fig. 7.2 An area pattern generated by the Johnson–Mehl model. Numbers
indicate nucleation times of nodes

Table 7.2. *Moments for properties of the pattern illustrated in fig. 7.2*

Property	Observed values	Expected values
$E(N)$	5.370	<6.000
$E(S)$	4.911	5.404
$E(L)$	0.868	<0.907
$E(A)$	1.577	2.094
$E(C)$	4.000	4.000

$\alpha = 0.08$ $v = 0.20$

ability for various properties or examine the correlation coefficients for relationships between pairs of properties. Table 7.3 gives estimates of the coefficients of variability and Pearsonian product–moment correlation coefficients for selected properties. The values were obtained from patterns produced from a series of simulation runs of the J–M model (Boots, 1975*b*).

We should note that the J–M model, unlike the cell model, is exclusively a growth model. In addition, it should be apparent that given a 'growth' interpretation the cell model can be considered a special case of the J–M model, one in which the time dimension is either negligible or non-existent. Consequently, if α or v, or both, are sufficiently large so that the entire plane is colonized in one time period we replicate the cell model.

To date, the J–M model has only been used occasionally in geography, although it is beginning to be extensively explored in biology (Glass, 1973; Armstrong, 1974; Jackson, 1974). Boots (1973) employed a modified J–M model as a standard in examining the patterns of bus service hinterlands in

Table 7.3. *Estimates of coefficients of variation and correlation coefficients for properties of the Johnson–Mehl model*

Coefficients of variation				
Property				
Contact number	(N)	0.497	(S.D. 0.113)	
Full neighbors	(C)	0.413	(S.D. 0.054)	
Area	(A)	0.824	(S.D. 0.194)	
Perimeter	(S)	0.540	(S.D. 0.092)	
Edge length	(L)	0.323	(S.D. 0.044)	

Correlation coefficients				
	C	A	S	L
N	0.796	0.817	0.840	0.242
C		0.765	0.747	0.311
A			0.946	0.580
S				0.678

S.D. = Standard deviation

the United Kingdom described previously (see pp. 132–5). The modification used was in terms of α, the nucleation rate. This was changed from one which was constant throughout the time dimension to one in which the number of new nuclei created in any time period was distributed according to a Poisson process. Consequently, in this modified J–M model the locations of the nuclei in space and in time were both Poisson distributed. In addition, unsuccessful nuclei (those that would be located in areas of the plane already occupied) were relocated in unoccupied portions of the plane. In general, good fits were obtained between the properties of the modified model and those of the empirical patterns (Boots, 1973).

Of course, other modifications can be made to the model. Potential modifications involve considerations of different ways of locating the nuclei and different ways in which the associated cells grow. These two circumstances are essentially controlled by the choice of α (the nucleation rate) and v (the lateral rate of growth of individual cells). In the basic model described above both of these values are assumed to be constant throughout the duration of the process, although after the elimination of unsuccessful nuclei, the true nucleation rate, α', actually decreased over time.

The nucleation rate, α has two component parts, one of which is responsible for the location of the nucleus in space (its A_i coordinate) and the other which is responsible for the nucleus's location in time (its t_i coordinate). In the basic model the A_i coordinates were selected at random subject to the constraint that locations selected in already assigned portions of the plane would lead to the immediate death of the nucleus. Perhaps the simplest modification of the model is to retain all of the conditions of the model but to select the A_i coordinates for the nuclei according to some process other than a Poisson one. Such an approach was used by Johnson and Mehl (1939) who presented an alternative form of their model in which nuclei could only be born at the boundaries of already existing cells. Along these lines, one modification which might have geographic implications is to ensure that locations chosen for new nuclei are at least a minimum distance from the closest point on the perimeter of existing cells. Such a constraint might be appropriate in the development of store trading areas or similar territories.

The most interesting, and to the geographer perhaps the most valuable, modifications of the model involve changes in the values of α and v. Consideration of such changes enables the incorporation of non-spatial system parameters into the model. Changes in α and v may be mutual or independent. Let us first consider changes in α. The nucleation rate can be changed in response to the occurrence of environmental factors which influence the establishment of nuclei. Favorable environmental conditions might lead to positive feedback in the system giving an impetus to the creation of new nuclei which might be met by an increase in the nucleation rate over time. Such a situation might occur in settlement in pioneer areas where the initial establishment of settlements is slow, but once a basic network has been established and the environment 'tamed' conditions facilitate an increase in the establishment of new settlements. It is also possible to imagine instances of negative feedback from the environment which could lead to a reduction in the nucleation rate over time (or in extreme cases a premature cessation in the location of new nuclei, or perhaps even the removal of existing nuclei giving a negative net nucleation rate). Such a situation might represent an alternative response to settlement in a 'pioneer environment'. If the environment is sufficiently 'hostile' fewer and fewer new settlements would be introduced while existing centers may find it extremely difficult to survive and may in fact disappear if conditions so dictate. In contrast, there are many instances in which simple 'overcrowding' could lead to a reduction in the nucleation rate. In fact, in the basic model the true nucleation rate, α', actually decreases over time as the plane becomes more crowded.

In a similar way, the individual cell growth rate, v, can be changed to a variable if real world circumstances suggest this is appropriate. It is not difficult to envisage circumstances in which this might occur. The growth rate of a market area may increase with an increase in the size of its area if economies of scale can be implemented in the production and distribution of goods. Similarly, a political area may benefit from an increase in physical size if this results in a diversification and increase in its resource base. In this respect, it should be noted that in situations of increasing rates of cell growth it is possible to create embedded cells not recognized in the operation of the basic model (Boots, 1972).

Instances in which one might expect a decreasing growth rate over time are not uncommon. There are many times when upper thresholds might be imposed on the size of cells beyond which further growth cannot occur. In socio-economic networks such thresholds are often the product of transportation and other technological constraints.

Of course, it is quite possible that both α and v will change over time in other than monotonic increasing or decreasing fashions. It is possible to simulate the J—M model allowing both α and v to be controlled by either internal or external parameters. Indeed, in the guise of a simulation model, the basic framework of the J—M model may prove especially useful as a model for evaluating particular empirical instances. Perhaps many of the instances displayed in fig. 1.3 (p. 6) could be modeled after territories around points are assumed to be characterized by a particular pattern.

8

Area patterns: clumping models

We now turn to the examination of models which incorporate processes leading to patterns of overlapping areas which may or may not be space exhaustive. So far only one basic model has been developed extensively but the nature of this model is such that it has given rise to a series of related approaches.

8.1.   The Roach approach
The basic model was developed to aid persons concerned with counting very small particles on planar surfaces where the problem of underestimating the count often arises because overlapping particles can be visually indistinguishable from isolated particles. Although several researchers have been active in developing this model, we first base our approach on the work of Roach (1968) who has incorporated much of the previous work of others into his own studies. Despite the simplicity of the model and the considerable research effort expended upon it, analytical results are difficult to derive and many of the results reported are only approximations of the true values.

The **P** assumptions locate the points in the plane according to a Poisson process of density λ points per unit area. The **A** assumption consists of considering each of the points generated by the **P** assumptions as the center of a circle of radius r. All points give rise to circles of the same radius. Some of the circles overlap and *clumps* are formed. Fig. 8.1 illustrates a pattern generated by the model. Clumps can be of various sizes and, in general, a clump of size n may be defined as a set of n circles each of which has at least one point in common with one or more of the other $(n - 1)$ circles. Any pair will overlap if the distance between them is less than the diameter of the circles. The reader will note this is essentially the same model as the circuit model we discussed in chapter 5 (pp. 104–7). In the circuit model a pair of points became linked if the distance between them was less than a critical distance. In the clumping model a pair of points become linked in a clump if the distance between them is less than a critical distance defined by $2r$.

152

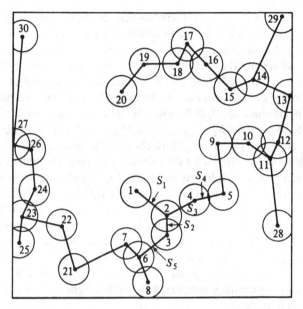

Fig. 8.1 An area pattern generated by the clumping model

The most interesting properties are the expected number of separate clumps produced by the process and the size of these clumps. Here we also discuss results derived for a modified model. The **A** assumption of equal radii for all circles is replaced by one which allows the individual circles to have different radii.

Roach (1968) transforms this problem into one of finding the frequency distribution of the distances between the set of points located in the plane by a Poisson process. Assume that the density of points generated by the **P** assumptions is λ points per unit area. One of the points is selected, arbitrarily as a starting point. This point is linked to the point closest to it. The distance between the linked points is labelled S_1 (see fig. 8.1). The locations of the remaining unlinked points are examined and the point which is nearest to either of the two points already joined is linked to whichever of those two points it is closest to. This new interval is labelled S_2. One of the remaining unlinked points is now nearer to one of the three already joined points than to any other and this point is linked to whichever of the three joined points it is closest to. The new interval is labelled S_3. This 'nearest neighbor' linkage procedure is continued indefinitely. In this way a sequence of intervals $S_1, S_2,$ $\ldots, S_{n-1}, S_n, S_{n+1}, \ldots$ is obtained. If the distance S_1 is less than $2r$ the two circles drawn about the points will overlap. Similarly, if S_2 is less than $2r$ the third circle will overlap at least one of the first two. If the successive intervals $S_1, S_2, \ldots, S_{n-1}$ are all less than $2r$ the n points will all be members of the same clump. If the next interval, S_n, is greater than $2r$, the original circle chosen at random will be a member of an isolated clump of size n.

If we wish to determine the probability that a circle is a constituent of a clump of size n we require the probability of the occurrence of a set of $(n-1)$ intervals, $S_1, S_2, \ldots, S_{n-1}$, all of which are less than $2r$, followed by an nth interval, $S_n \geqslant 2r$. We can consider each of the intervals, S_i, as an outcome of a trial, a success occurring when $S_i \geqslant 2r$, and a failure occurring when $S_i < 2r$.

The sequence of distances $S_1, S_2, \ldots, S_n$ can be considered as a series of such trials. For any given pattern of points, the set of intervals $S_1, S_2, \ldots, S_n$ is unique. However, by choosing a different starting point the order of these intervals can be changed. In this way the size of the nth interval will be virtually independent of the value of n. The probability of success on the nth trial will only be slightly different from the probability on any of the other $(n-1)$ trials. This approximation offers a method of estimating the required values. If this approximation was unavailable the solution would necessitate evaluating all the possible combinations for all possible sizes of clumps. This would provide the exact probability, but Roach has demonstrated that for $n=4$ the geometry is extremely complex, and that for higher values of n the solution becomes intractable.

We begin by determining the probability, p_1, that the center of a circle, C_i, located by a Poisson process in the plane is isolated. This is the same as the probability that no circle center C_j ($j \neq i$) lies within a circle of radius $2r$ of C_i. Thus,

$$p_1 = \exp\left[-\pi(2r)^2\lambda\right] \tag{8.1}$$

The probability, p_n, that a circle is part of an isolated clump of size n is

$$p_n \simeq (1-p_1)^{n-1}p_1$$

$$= [1 - \exp(-4\pi r^2\lambda)]^{n-1}[\exp(-4\pi r^2\lambda)] \tag{8.2}$$

If the mean number of circles per unit area is λ, the mean number of clumps of size n per unit area, C_n, is given by

$$\frac{p_n\lambda}{n} \simeq \frac{(1-p_1)^{n-1}p_1\lambda}{n} \tag{8.3}$$

and hence the mean number of clumps of all sizes, C, is obtained by summing expression (8.3) over all values of C_n.

$$C \simeq \lambda p_1 \left[1 - (1-p_1)/2 + ((1-p_1)^2)/3 + \ldots\right]$$

$$= \frac{\lambda p_1 \ln p_1}{p_1 - 1} \tag{8.4}$$

Substituting the value of p_1 from (8.1) into (8.4) we get

$$C \simeq \frac{4\pi r^2 \lambda^2}{\exp(4\pi r^2 \lambda) - 1} \tag{8.5}$$

To illustrate the utility of these results consider fig. 8.1. Thirty circles have been generated using the assumptions of the clumping model. The size of the study area is 90×90 units, so that $\lambda = 30/8100 = 0.0037$. The radius of each circle is 5 units. Expression (8.1) is used to obtain the expected proportion of isolated circles. The value derived is substituted into (8.2) to yield the expected number of circles in clumps of size two, three, etc. The values obtained are

Table 8.1. *Frequency distribution of clump sizes for the pattern illustrated in fig. 8.1*

(1) Size of clump	(2) Observed frequency	(1) x (2) Observed number of circles in each clump size	Expected number of circles in each clump size
1	10	10	9.381
2	4	8	6.447
$\geqslant 3$	4	12	14.172
(N)	18	30	30

$X^2 = 0.748$ $df = 1$ $\chi^2_{0.20} = 1.642$

given in table 8.1 together with the values obtained from the pattern in fig. 8.1. A X^2 test gives a calculated value of $X^2 = 0.748$ $(df = 1)$ so that we are in a position to accept the model in this instance. In addition, (8.5) gives the expected number of clumps per unit area as 0.0020 while the observed value is $18/8100 = 0.0022$.

Using a similar approach the model can be developed to cover those instances where the circles are not all of the same size. If the frequency distribution for the size of the radii of the circles, $f(R_1)$, is known, the expected number of circles whose radius lies in the interval between R_1 and $(R_1 + dR_1)$ is $f(R_1)dR_1$. The probability that a circle of radius R is not overlapped by any circle of a radius between R_1 and $(R_1 + dR_1)$ is

$$\exp\left[-\pi(R + R_1)^2 \lambda f(R_1)dR_1\right] \tag{8.6}$$

Thus, the probability that a circle of radius R is not overlapped by any other circle of any size is

$$\exp\left[-\pi\lambda \int_0^\infty (R + R_1)^2 f(R_1)dR_1\right]$$

$$= \exp\left[-\pi\lambda(R^2 + 2\mu R + \mu^2 + \sigma^2)\right] \tag{8.7}$$

where μ = mean of $f(R_1)$ and σ^2 = variance of $f(R_1)$.

The probability that any circle is isolated, p_1, is

$$p_1 = \int_0^\infty f(R) \exp\left[-\pi\lambda(R^2 + 2\mu R + \mu^2 + \sigma^2)\right] dR \tag{8.8}$$

If the expression in (8.8) proves difficult or impossible to integrate, p_1 may be approximated as follows

$$p_1 \simeq \sum_i p_i \exp\left[-\pi\lambda(R^2 + 2\mu R + \mu^2 + \sigma^2)\right] \tag{8.9}$$

where p_i = proportion of circles in a given size interval i. The values obtained from expressions (8.8) and (8.9) can be substituted in equations (8.3) and (8.5) to obtain the mean number of clumps of size n, C_n, and the mean number of clumps of all sizes, C.

8.2.   The Getis–Jackson model

Getis and Jackson (1971) have also developed a clumping model but their investigation is concerned with answering a different question. Rather than determine how many clumps of a given size exist, they ask what is the probability that an infinitely small area of the plane is covered by k overlapping circles (where $k = 0, 1, \ldots, n$). For the basic model in which points are distributed according to a Poisson process of density λ, the probability, p_k, that an infinitesimal area, dA, located at a point C, is covered by k circles is the probability that centers of k circles lie within a radius r of C. This is given by the Poisson expression

$$P(k; \mu) = (\mu^k e^{-\mu})/k! \tag{8.10}$$

where $\mu = \pi r^2 \lambda$.

Even though the probabilities from different infinitesimal areas dA are not independent, the expected portion of the whole plane covered by exactly k sources is given by (8.10). Using (8.10) we can derive the expected proportions of the study area in fig. 8.1 covered by none, one, two, etc. circles. Table 8.2

Table 8.2. *Proportion of study area covered by k circles (based on fig. 8.1)*

k	Observed proportion of study area covered by k circles	Expected proportion of study area covered by k circles
0	0.7443	0.7478
1	0.2372	0.2173
2	0.0183	0.0316
$\geqslant 3$	0.0002	0.0033

gives the values so derived together with the observed values. Evidently, there is a good fit between the two sets of values.

This approach can be extended to the version of the model in which the radii of the circles are of different sizes. Assume that the radii have a discrete frequency distribution in which the radius, r, can take on a set of values, r_i, with probability p_i. This is the same as if, for each index i, the sources of radius r_i were distributed in a Poisson manner with rate λp_i, independently of the location of circles with different radii. The probability that k_i circles of radius r_i cover an infinitesimal area, dA, at C is the probability that k_i such points lie within a radius r_i of C, and so is given by (8.10) in which k_i replaces k and $\mu_i = \pi r_i^2 \lambda p_i$ replaces μ. The total number of circles which cover dA is $k = \sum_i k_i$ and since the k_i circles are independent, k has a Poisson distribution with parameter

$$\mu = \sum_i \mu_i = \pi\lambda \sum_i r_i^2 p_i = \pi\lambda E(r^2) \tag{8.11}$$

where $E(r^2)$ = expectation of r^2 with respect to the distribution of r. The proportion of the plane covered by k circles is given by (8.10) with μ taking the value obtained from (8.11). This argument can be extended to cover continuous distributions of r. Again, all that is required is the substitution of the appropriate value of $E(r^2)$ in (8.11).

The basic model is useful in the analysis of patterns of phenomena which can be represented as circles. Lee (1972) used the Roach approach to analyze patterns of urban settlement and urban spheres of influence while the work of Getis and Jackson (1971) was originally developed in order to study the expected proportion of a study area polluted by a given number of Poisson distributed point sources. Subsequently, Getis and Jackson extended their model to handle clumps of shapes other than circles and an application of their results is discussed below.

Lee's study of the clumping of cities involves a slight re-orientation of the model. Since urban settlements do not overlap physically or politically, cities are considered to form clumps if any parts of their boundaries are contiguous. Lee's study area comprised discontinuous portions of approximately one-third of the continental United States. He suggests that cities can be represented as circles of various sizes, the radii of which are proportional to the population of the cities. Using the results derived by Nordbeck (1965) for the relationship between city size and population, Lee is able to convert the cities' population sizes into radii. In this way he is able to obtain a size frequency equation for urban settlements in his study area. This is

$$f(r) = 0.86439 r^{-2.895} \qquad (1.10176 \leqslant r \leqslant 15.38155)$$

Table 8.3. *Frequency distribution of clump sizes for pattern of cities*

Size of clump (x)	Observed frequency	Observed total in clumps of size x	Expected total in clumps of size x
1	1152	1152	1218.10
2	45	90	94.03
3	7	21	7.26 ⎫
4	2	8	0.56 ⎬
5	2	10 ⎫	⎭
$\geqslant 6 \begin{bmatrix} 7 \\ 10 \\ 15 \end{bmatrix}$	4	39 ⎬⎭	0.05 ⎭
(N)	1211	1320	1320

$X^2 = 628.9$ $df = 1$ $\chi^2_{0.20} = 1.642$

Calculated from data given in Lee (1972).

The values for r are in kilometers and the limits of r are equivalent to the imposition of upper and lower population limits of 1 000 000 and 2500, respectively. The values of $f(r)$, μ, and σ^2 are substituted in (8.9) to yield an estimated value of p_1 which in turn is used to give C_n and C. The results are given in table 8.3. A X^2 test applied to this data indicates that we must reject the model ($X^2 = 628.89, df = 2$).

Lee also used the same procedure to examine the clumping of urban spheres of influence defined by 50% or more commuting fields of central cities. Unlike the cities, commuting fields truly overlap. The size frequency equation of the fields was estimated as

$$c(r) = 5.667508r^{-3.104555} \qquad [2.57 \leqslant r \leqslant 56.64 \text{ (km)}]$$

where $c(r)$ = number of commuting fields, and r = radius of circle representing area of commuting field. Using this equation, the expected number of clumps can be found and is given in table 8.4.

Table 8.4. *Frequency distribution of clump sizes for pattern of urban spheres of influence*

Size of clump (x)	Observed frequency	Observed total in clumps of size x	Expected total in clumps of size x
1	380	380	381.09
2	5	10	11.55
$\geqslant 3$	1	3	0.36
(N)	386	393	393

Calculated from data given in Lee (1972).

Although the model appears to hold up, Lee suggests caution in interpreting the results because of data collection problems and the fact that many large cities are bordered by a large number of suburban settlements. In addition, Lee did not assume that the locations of the urban settlements followed a Poisson-generated pattern. It is quite possible that the locations of settlements in his study area are significantly different from those expected under the conditions of a Poisson process as assumed in the model. Table 8.4, however, shows that a good fit is obtained for the commuting fields.

It is possible to generate a whole series of related models by changing either the **P** assumptions, the **A** assumptions, or both sets of assumptions of the basic clumping model. Changes in the **P** assumptions would involve the use of distributions other than the Poisson process for locating the points which become the centers of the areas created by the **A** assumptions. So far no efforts have been made in this direction.

Changes in the **A** assumptions can entail the generation of shapes other than circles about the points located by the **P** assumptions. Roach (1968) considers the effect if ellipses or squares replace the circles of the basic model. In earlier work, Mack (1954, 1956) demonstrated that even when the areas created by the **A** assumptions are of different shapes and sizes, only the size and the perimeter length of the areas are relevant to the determination of the number of clumps. The shapes of the areas are immaterial. This is a general result which is true subject only to the following three conditions:

(i) all the areas are convex (although they need not be symmetrical);
(ii) all the areas are oriented with no preferred direction (i.e. as though a table of random numbers were used for the orientation process);
(iii) there are a large number of each kind of area.

Mack (1954) found that in a unit area of the plane the expected number of isolated areas, C_1, is given by

$$C_1 = \sum_r N_r \exp\left[-\sum_\mu N_u\left(a_u + a_r + \frac{S_u S_r}{2\pi}\right)\right] \qquad (8.12)$$

where N_r = number of areas of size a_r and perimeter length S_r in a unit area of the plane.

In their alternative approach to the clumping model, Getis and Jackson (1971) also consider shapes other than circles for the areas generated by the **A** assumptions. If the areas are identical in shape, size, and orientation, and have an area A, the proportion of the entire plane covered by k areas is still the same as that in (8.10) except that $A\lambda$ replaces $\pi r^2 \lambda$ as the value of μ. Expression (8.10) may also be used when the areas are similar figures of the same orientation. If we let A denote the area of one of the figures chosen as a basis for comparison, and let r denote the ratio of linear dimensions of any other figure to the basic one (giving an area r^2A), the value of μ in (8.10) is

$$\mu = A\lambda E(r^2) \tag{8.13}$$

which when substituted in (8.10) gives

$$P(k; A, \lambda, E(r^2)) = \frac{[A\lambda E(r^2)]^k \exp[-A\lambda E(r^2)]}{k!} \tag{8.14}$$

It was this version of the model which Boots *et al.* (1972) used to examine the distribution of sulfur oxide pollution (SO_x) from point sources over the northeast New Jersey – New York City region of the United States. Thirty-seven major point sources of pollution located in the area of greatest SO_x concentrations were examined. The locations of these sources were found to be not significantly different from those expected from a Poisson process. A pollution zone was associated with each of these point sources. A zone was defined by the boundaries enclosing an area exhibiting some specified amount of air pollution. The study considered zones including 0.01 parts per million (p.p.m.) and greater SO_x emanating from a point source. As a simplifying condition it was assumed that within each pollution zone associated with a particular source the amount of pollution is everywhere the same. The size of each zone was considered to be a function of the amount of pollution originating at the point source. Ellipses were chosen to represent the zones and their orientation was determined by prevailing meteorological conditions. The model was tested for summer conditions in order to minimize the amount of SO_x from space-heating sources. The zone sizes were assumed to be exponentially distributed. In this case the value for $E(r^2)$ is $2b^{-2}$ and (8.14) becomes

$$P(k) = \frac{[\pi\lambda(2/b^2)]^k \exp[-\pi\lambda(2/b^2)]}{k!} \tag{8.15}$$

where $1/b$ is the mean of the exponential distribution of the form be^{-br}.

Table 8.5 gives the expected proportion of the study area covered by $k = 0$, 1, 2, ... overlapping pollution zones. The expectation for no pollution is about one per cent of the total area. Following the assumption that pollution rates are everywhere the same within a zone associated with a point source, a value of $k = 1$ is equivalent to the 0.01 p.p.m. level of pollution, $k = 2$ implies the 0.02 p.p.m. level and so on. As table 8.5 shows the expected values were compared with two sets of observations. The first, (a), is a frequency distribution based on SO_x levels at each of twenty-three measuring stations. The second, (b), is a measure of the proportion of the study area covered by various levels of pollution on an average day. Using a K–S test for comparability the model cannot be rejected at the 0.20 level for the station data, but is rejected for the area data.

Table 8.5. *Observed and expected proportions of study area covered by k pollution zones*

Number of overlapping air pollution regions (k)	SO_x (in p.p.m.)	Expected proportion of area covered by k	Observed proportions		Cumulative proportions		
			By stations (a)	By areas (b)	Expected	(a)	(b)
0	0.000–0.009	0.01	0.00	0.00	0.01	0.00	0.00
1	0.010–0.019	0.06	0.00	0.00	0.07	0.00	0.00
2	0.020–0.029	0.13	0.04	0.05	0.20	0.04	0.05
3	0.030–0.039	0.18	0.17	0.09	0.38	0.21*	0.14**
4	0.040–0.049	0.19	0.22	0.31	0.57	0.43	0.45
5	0.050–0.059	0.17	0.26	0.36	0.74	0.69	0.81
6	0.060–0.069	0.12	0.17	0.08	0.86	0.86	0.89
7	0.070–0.079	0.07	0.09	0.06	0.93	0.95	0.95
8	0.080–0.089	0.04	0.04	0.05	0.97	0.99	1.00
≥9	≥0.090	0.03	0.00	0.00	1.00	1.00	1.00

* $D_{obs} = 0.17$ ** $D_{obs} = 0.24$ $D_{0.20} = 0.18$

Modified from Boots, Getis and Hagevik (1972, table 1, p. 384)

Along somewhat different lines another possible modification is the addition of a time dimension to the model. In view of the properties examined with relation to the basic model it was immaterial whether we considered the points resulting from the **P** assumptions to be of simultaneous origin or whether we felt they were created in a sequential fashion. A somewhat different emphasis can be put on the model if we assume that the points are located sequentially such that each point possesses a location in time as well as in space. If, in addition, we assume that the circle (or other shape) associated with each nucleus will obliterate all existing nuclei and parts of existing cells which lie within its range (as shown in fig. 8.2), two additional important

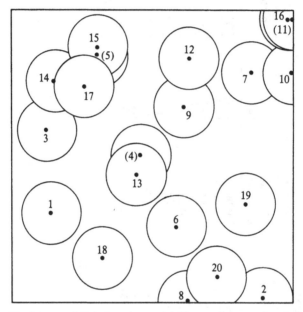

Fig. 8.2 A modified clumping model. The numbers associated with the nuclei represent the order in which they were generated. Numbers in parentheses indicate nuclei which were obliterated by the areas associated with nuclei established subsequently

questions can be posed. First, what is the distribution of nuclei that will be observed at any time, t_i? Second, how many undisturbed circles (i.e. circles for which the entire perimeter can be traced without interruption) exist at any time, t_i? Such a modified model might be appropriate in some instances of spatial competition, e.g. store trade areas where new competitors are introduced into the system through time. Modifications along these lines have already been undertaken by some astronomers in an attempt to explain the observed distribution of impact craters on the lunar surface. In a series of articles Marcus (1966, 1967), Fielder (1966), and Fielder and Marcus (1967) have explored several aspects of the two questions posed above. Using a computer simulation, Cross and Fisher (1968) examined the model for instances where the number of circles of a given radius, R, was inversely proportional to R^2. Circles (craters) were located following a Poisson process in two dimensions and occurred at random intervals in time. In many instances results similar to those obtained from photographs of the moon's surface were obtained.

Appendix A:

Introduction to probability theory

In the study of point patterns, statements about the likelihood of an outcome in a given situation are needed. The central question is: how probable is a particular map pattern given the presumed existence of a particular spatial process? In chapter 1 we define what is meant by spatial process. In this appendix we briefly discuss the notion of probability. The goals are (1) to make clear how outcomes of experiments can be enumerated and (2) to show how probabilities can be attached to each possible outcome. Since explanations are in terms of set notation, the discussion begins with an introduction to the study of sets. A more advanced reader may skim this appendix noting the aspects of probability theory that are emphasized.

A.1. Set notation

In point pattern analysis, an *experiment* is considered to be as simple as noting the number of points in a map region or measuring the distance between two points. The result of an experiment is called an *outcome*. It is usually the case that before an experiment is performed all possible outcomes are identified. These outcomes become the *elements* of a set. Of course, for some sets the number of elements would be too large to list.

A set can contain all possible outcomes of an experiment or just some of them. When a set contains all possible outcomes it is called a sample space. If the number of possible outcomes is countable then these are members of *discrete sample spaces* (as opposed to continuous sample spaces). We will confine our attention to discrete sample spaces.

Venn diagrams aid in visualizing operations on sets. These are rectangles or other convenient shapes which represent all possible outcomes of an experiment (see fig. A.1). Circles drawn within the rectangle represent *subsets*, i.e. some portion of all outcomes. Let S represent the sample space and A and B represent subsets. Suppose that S contains $\{1, 2, 3, 4, 5, 6\}$, A contains $\{1, 2, 3\}$ and B contains $\{2, 3, 4\}$, then the subsets overlap as shown in fig. A.2. The shaded area represents the elements of S which are common to A and B. Sym-

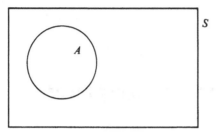

Fig. A.1 Venn diagram containing subset *A*

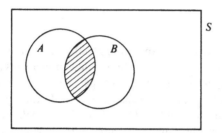

Fig. A.2 Venn diagram showing overlapping subsets

bolically the elements of *A* and *B* are represented as $A \cap B$. In this case $A \cap B$ = {2, 3}. The symbol $\cap$ reads 'intersection', 'and' or 'cap'.

The sum total of all outcomes contained in *A* or *B* is {1, 2, 3, 4}. Thus $A \cup B$ represents all the different outcomes in *A* and *B* taken together. The symbol $\cup$ reads 'union', 'or', 'cup'. Fig. A.3 summarizes this discussion.

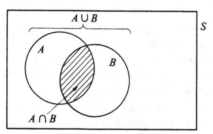

Fig. A.3 Venn diagram showing an intersection and a union of subsets

The manipulation of these concepts mathematically is called the algebra of sets. It is not necessary for us to develop this subject; it is well that the terms 'experiment', 'outcome', 'sample space', 'set', 'subset', 'intersection' and 'union' are clear.

A.2.   Combinatorial theory

In this section various methods for enumerating outcomes are discussed. Here we show ways of listing all possible outcomes of an experiment.

In the section which follows probabilities are attached to each of the listed outcomes.

Suppose there are three sets A_1, A_2 and A_3 each containing a group of points. How many different ways can one point from each set be selected? A_1 contains $n_1 = 3$ red points labeled r_1, r_2, r_3; A_2 contains $n_2 = 5$ green points labeled g_1, g_2, g_3, g_4, g_5; and A_3 contains $n_3 = 2$ blue points labeled b_1, b_2. If an r_1 is selected first then we have a choice of 5 green points. Of the five green points suppose we select g_1 and we have two remaining choices for an element from A_3. The different ways we can select one element from each set are listed below:

$r_1g_1b_1$	$r_1g_1b_2$
$r_1g_2b_1$	$r_1g_2b_2$
$r_1g_3b_1$	$r_1g_3b_2$
$r_1g_4b_1$	$r_1g_4b_2$
$r_1g_5b_1$	$r_1g_5b_2$
$r_2g_1b_1$	$r_2g_1b_2$
$r_2g_2b_1$	$r_2g_2b_2$
$r_2g_3b_1$	$r_2g_3b_2$
$r_2g_4b_1$	$r_2g_4b_2$
$r_2g_5b_1$	$r_2g_5b_2$
$r_3g_1b_1$	$r_3g_1b_2$
$r_3g_2b_1$	$r_3g_2b_2$
$r_3g_3b_1$	$r_3g_3b_2$
$r_3g_4b_1$	$r_3g_4b_2$
$r_3g_5b_1$	$r_3g_5b_2$

This same selection process can be represented in a tree diagram (fig. A.4), where each branch beginning at point 0 represents another arrangement. Each line intersection represents a decision point, i.e. a place where a choice is made. The darker line represents the sequence $r_2g_1b_2$.

The list and the tree diagram are two ways of showing all the outcomes of an experiment, but in most studies it is only the total number of possible outcomes that is important. In the above example, the number of possible different selections is 30. This is equivalent to $n_1 \cdot n_2 \cdot n_3 = (3)(5)(2) = 30$. Thus, if sets $A_1, A_2, \ldots, A_k$ have, respectively, $n_1, n_2, \ldots, n_k$ elements, there are $n_1 \cdot n_2 \cdot \ldots \cdot n_k$ different ways in which one can first select an element of A_1, then an element of $A_2, \ldots$, and finally an element of A_k. For example, on a map having four subareas, six cities, and fifteen routes connecting the cities there are $4 \cdot 6 \cdot 15 = 360$ different ways in which one can choose a subarea, a city, and a route.

A.2.1. Permutations

Suppose we want to know the number of different ways of selecting three of the six cities mentioned above (label them A, B, C, D, E, F). If it

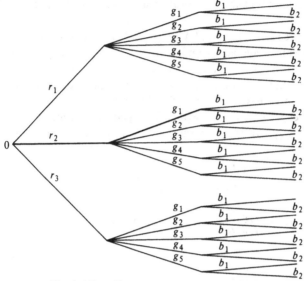

Fig. A.4 Tree diagram

matters what the order of choice is then *ABC* is different from *BAC* even
though the same cities are considered. There are six different ways one can
select the first city. Having chosen one of these there are five remaining choices
for the second city. There are four possibilities for the third city. This procedure can be written as $6 \cdot 5 \cdot 4 = 120$ or the permutation of three objects
selected from a set of six objects is 120.

A simplified way of writing a general formula for permutations is with
factorial notation. Here $n \cdot (n-1) \cdot (n-2) \cdot \ldots \cdot 3 \cdot 2 \cdot 1$ is $n!$ or *n factorial*.
By definition $0! = 1$. If $n!$ is divided by $(n-x)!$, the result is $n(n-1)(n-2)$
$\ldots (n-x+1)$. Take the example just mentioned.

We have

$$\frac{n!}{(n-x)!} = \frac{6!}{3!} = 6 \cdot 5 \cdot 4 = 120 \tag{A.1}$$

A.2.2. Combinations

For most problems dealt with in this book there is no concern for
the order in which objects are selected. In the case of the three cities mentioned
above it usually matters more which three cities are chosen rather than the
order in which they are selected. *Combinations* are the number of unordered
selections. The number of combinations is always less than or equal to the
number of permutations. The question posed is: what is the number of unordered ways three (x) of the six cities (n) may be selected? The twenty

different ways are listed below.

ABC	*BCD*
ABD	*BCE*
ABE	*BCF*
ABF	*BDE*
ACD	*BDF*
ACE	*BEF*
ACF	*CDE*
ADE	*CDF*
ADF	*CEF*
AEF	*DEF*

Notice that no combination of three letters is repeated. The first combination, *ABC*, can be permuted in six different ways (*ABC, ACB, BAC, BCA, CBA, CAB*). In this example there is one-sixth as many combinations as there are permutations. If the total number of permutations is divided by $x!$ we get the number of combinations. This is:

$$\frac{n!}{x!(n-x)!} = \frac{6!}{3!\,3!} = 20 \text{ combinations} \tag{A.2}$$

The above formula is read: the number of combinations of x objects selected from a set of n objects. It is often denoted as $\binom{n}{x}$.

Tables of factorials are generally available and on some electronic calculators factorials can be evaluated directly. One can also use Stirling's formula which approximates $n!$ when n is large. This is

$$\sqrt{(2\pi n)}\left(\frac{n}{e}\right)^n \tag{A.3}$$

where e, the base of natural logarithms, is approximately 2.7183. Suppose $n = 40$, Stirling's formula gives $(8.142)^{47}$ while $40! = (8.159)^{47}$.

A.3.   Binomial coefficients

A polynomial made up of two terms, such as $y + z$, is called a binomial. The coefficients of the variables in the expansion of $(y + z)^n$ are called binomial coefficients. The combinatorial value $\binom{n}{x}$ given earlier is a binomial coefficient. Why is this the case?

If n is a positive integer and we multiply out $(y + z)^n$ term by term for $n = 3$ we get

$$\begin{aligned}
(y + z)^n &= (y + z)(y + z)(y + z) \\
&= y \cdot y \cdot y + y \cdot y \cdot z + z \cdot z \cdot y + z \cdot y \cdot y + y \cdot z \cdot z + \\
&\quad z \cdot y \cdot y + z \cdot z \cdot y + z \cdot z \cdot z \\
&= y^3 + 3y^2z + 3yz^2 + z^3
\end{aligned} \tag{A.4}$$

Note that the coefficient of y^2z is three, because there are three ways in which one z and two y's can be selected, one from each of the three factors $(y + z)$. Similarly, the coefficient of yz^2 is three because there are three ways in which two z's and one y can be selected. The coefficients of y^3 and z^3 are both equal to one because there is only one way in which three y's or three z's can be selected, one from each of the three factors of $(y + z)$.

Thus it can be seen that to get the first term, y^3, we select no z's and three y's from the three factors. Because this can be done in $\binom{3}{0}$ or one way, y^3 occurs only once in the expansion and its coefficient is one. The second term can be selected in $\binom{3}{1}$ or three ways and so on. The complete expansion is the sum of four terms:

$$(y + z)^3 = \binom{3}{0} y^3 + \binom{3}{1} y^2 z + \binom{3}{2} yz^2 + \binom{3}{3} z^3$$

The general expression is

$$(y + z)^n = \binom{n}{0} y^n + \binom{n}{1} y^{n-1} z + \binom{n}{2} y^{n-2} z^2 + \ldots + \binom{n}{x} y^{n-x} z^x +$$

$$\ldots + \binom{n}{n} z^n$$

$$= \sum_{x=0}^{n} \binom{n}{x} y^{n-x} z^x \qquad (A.5)$$

A.4.   Multinomial coefficient

Suppose a set of n objects consists of x_1 which are alike, x_2 others which are alike, $x_3, x_4, \ldots, x_k$ others which are alike. The total number of different arrangements is

$$\frac{n!}{x_1! x_2! \ldots x_k!} \qquad (A.6)$$

As an example let five points be placed along a line where the line represents a highway. One of the points represents a city (C), two represent towns (T), and two villages (V). The possible arrangements are

$$\frac{5!}{1!\, 2!\, 2!} = 30$$

They are listed below.

CTTVV	*TCTVV*	*TTCVV*	*TTVCV*	*TTVVC*
CTVTV	*TCVTV*	*TVCTV*	*TVTCV*	*TVTVC*
CTVVT	*TCVVT*	*TVCVT*	*TVVCT*	*TVVTC*
CVTVT	*VCTTV*	*VTCVT*	*VTTCV*	*VTTVC*
CVVTT	*VCVTT*	*VVCTT*	*VVTCT*	*VVTTC*
CVTTV	*VCTVT*	*VTCTV*	*VTVCT*	*VTVTC*

A.5.   Occupancy theory

Here we are concerned with the placement of points into areas or cells. The question that is often asked is what are the number of ways points can be distributed among a given number of cells.

Suppose that there are four points to be placed in two cells. One way would be $\cdots | \cdot$ where the vertical line represents the boundary or partition between the two cells. Note that the vertical line can be drawn in five different places relative to the points $- | \cdots , \cdot | \cdots , \cdot \cdot | \cdot \cdot , \cdots | \cdot , \cdots | $. Thus, the total number of ways we can arrange the four points and the one partition is five, namely $\binom{5}{1}$ = 5. A general formula in which x points can be distributed among n cells is

$$\binom{x+n-1}{n-1} = \frac{(x+n-1)!}{(n-1)!\,x!} \tag{A.7}$$

For ten points in five cells, the occupancy result is 1001.

The theory explains such partition results under widely varying conditions such as restrictive cases where limits are placed on the number of points accepted into a cell and where the points can be differentiated according to some criterion.

A.6.   Probability distributions

Now that we have introduced ways to enumerate possible outcomes, we will turn to the second purpose of this appendix, i.e. attaching probabilities to the possible outcomes.

The classic demonstration device, a pair of dice, is used to illustrate that each possible outcome of a dice throw (2, 3, 4, 5, 6, 7, 8, 9, 10, 11, 12) is not equally likely. Intuitively, we know that the total 2 or 12 is less apt to occur than a 7. A 7 results from either rolls of 1 and 6, 2 and 5, 3 and 4, 6 and 1, 5 and 2, or 4 and 3 (or 6 different equally likely ways) while a 2 results from only the roll of a 1 and a 1. Table A.1 gives the number of ways each possible outcome occurs. Assuming that each of the 36 possible outcomes has a probability of 1/36, we can see that the probability associated with the number 7 is 6/36 or 0.167 while for the number 2 it is 1/36 or 0.028. Thus we might expect that about 17% of the time the value 7 will turn up and about 3% of the time a 2 will result.

By assigning a value to each possible outcome we create a *random variable*. The random variable is defined by the numbers representing the possible outcomes. In the dice case above, we have described a discrete rather than a continuous random variable, one which is defined in terms of a finite number of outcomes. The discussion which follows will be limited to discrete random variables. Some examples of random variables are: the number of heads in n flips of a coin, the number of spades held in a hand of bridge, the number of points in a cluster of points, the number of points in a cell.

Table A.1. *Outcomes for rolls of a pair of dice*

Possible outcome	Number of outcomes	Possible dice roll results	Probability
2	1	(1,1)	0.0278
3	2	(1,2; 2,1)	0.0556
4	3	(1,3; 3,1; 2,2)	0.0833
5	4	(1,4; 4,1; 3,2; 2,3)	0.1111
6	5	(1,5; 5,1; 4,2; 2,4; 3,3)	0.1389
7	6	(1,6; 6,1; 5,2; 2,5; 4,3; 3,4)	0.1667
8	5	(2,6; 6,2; 5,3; 3,5; 4,4)	0.1389
9	4	(3,6; 6,3; 5,4; 4,5)	0.1111
10	3	(4,6; 6,4; 5,5)	0.0833
11	2	(5,6; 6,5)	0.0556
12	1	(6,6)	0.0278
	36		1.0001

For the dice throw mentioned above, x, the discrete random variable, represents the integers 2 through 12. A formula can be developed which associates probabilities with outcomes, and this is the *probability function*, $P(x)$. In the case of the dice roll, the probability function is

$$P(x) = \frac{6 - |x - 7|}{36} \quad \text{for } x = 2, 3, \ldots, 11, 12 \qquad (A.8)$$

where the vertical bars enclose an absolute value which is either $(x - 7)$ or $(7 - x)$ whichever is positive, or zero. So, if x is 4, the probability assigned to $P(x)$ is

$$\frac{6 - |4 - 7|}{36} = \frac{6 - 3}{36} = \frac{3}{36}$$

which corresponds to the probability listed in table A.1. The entire set of probabilities for a random variable is called a *probability distribution*. Thus the right hand column in table A.1 represents the probability distribution, the total of which always sums to 1.00 (a rounding error of 0.0001 is evident).

In certain circumstances it is important to know the probability that x takes a particular value given that another random variable, y, also takes a particular value. The probabilities $P(x|y)$ are values of the conditional probability of the random variable x given that y takes on a particular value. For example, suppose random variables x, y represent the number of red points in square areas on a map and the total number of points in square areas, respectively. The answer to the following question is a conditional probability. What is the probability that three red points are observed in a square area which contains a total of five points?

In summary then, a probability function attaches a probability to a random variable and the entire set of probabilities for all values of the random variable is the probability distribution. The way in which outcomes of experiments are determined helps in developing the probability function and probability distribution. The entire procedure is summarized in fig. A.6 on page 177.

There are many probability distributions, some of which are so important that special names are given to them. We have selected several for discussion, mainly because they play a key role in the analysis of spatial patterns. Also, by discussing these distributions we will give additional examples of the way probabilities are attached to possible outcomes of experiments.

A.6.1. The binomial distribution

Earlier it was said that a binomial was a polynomial of two terms y and z. The y and z can stand for two possible outcomes of an experiment. If we know the probability of each particular outcome then by the expansion process shown earlier we can create a probability distribution for a series of experiments.

The binomial distribution aids us in determining the probability of obtaining x number of successes in a set of similar experiments. This is the dichotomous situation mentioned earlier. One of the two possible outcomes of a single binomial experiment is called a 'success' — the other is a 'failure'. Each of a certain number of heads observed in a coin-tossing experiment can be designated a 'success', as can each of a certain number of points found in a region. The binomial distribution is only applicable if the chance for a success is constant no matter how many experiments there are and if no one outcome is in any way affected by any other outcome (the assumption of independence).

To derive a formula for the binomial probability function we need only look back at the expansion of $(y + z)^n$ given earlier. In this case, however, the two terms are the probabilities of a success or a failure in any *one* of the experiments and the n represents the number of experiments. Let one term be called θ, the probability of a success, and the other term be $1 - \theta$, the probability of a failure. Note that $\theta + (1 - \theta) = 1$ which means that the results of any one binomial experiment is exhausted by the two possible outcomes, success or failure.

Let us assume that in a series of five independent experiments (e.g. coin tossing) the outcomes are success, success, failure, success, failure. The probability that this series of three successes and two failures would occur is $\theta^3(1 - \theta)^2$. But the combination of three successes and two failures could have occurred in $\binom{5}{3} = 10$ different equally likely ways. Hence the probability of x successes in n experiments is given by

$$P(x; n, \theta) = \binom{n}{x}\theta^x(1-\theta)^{n-x} \quad \text{for } x = 0, 1, 2, \ldots, n \qquad \text{(A.9)}$$

The term $P(x; n, \theta)$ means the probability function P depends upon the parameters n, the number of experiments, and θ, the probability of a success.

The x takes on whatever integer value required as long as the value of x is equal to or between 0 and n. The resulting distribution of probabilities is called the binomial distribution.

To further illustrate the use of (A.9), suppose that there are seven cells (regions). Each cell may be a recipient of one point. The probability of a cell receiving a point is assumed to be 0.6. What is the probability that five cells contain a point and two remain empty? This is the same as asking what is the probability of five successes (cells with a point) and two failures (cells without a point) in seven trials where the probability of a success is 0.6. This is

$$P(5; 7, 0.6) = \binom{7}{5}(0.6)^5(1 - 0.6)^2 = 0.261$$

Note that the binomial coefficient gives us 21 ways of arranging the five successes and two failures. Table A.2 lists the results for the other possible

Table A.2. *Binomial distribution for n = 7 and θ = 0.6*

Successes (x)	Binomial distribution	Probability
0	$\binom{7}{0} \cdot 6^0 \cdot 4^7$	0.002
1	$\binom{7}{1} \cdot 6^1 \cdot 4^6$	0.017
2	$\binom{7}{2} \cdot 6^2 \cdot 4^5$	0.077
3	$\binom{7}{3} \cdot 6^3 \cdot 4^4$	0.194
4	$\binom{7}{4} \cdot 6^4 \cdot 4^3$	0.290
5	$\binom{7}{5} \cdot 6^5 \cdot 4^2$	0.261
6	$\binom{7}{6} \cdot 6^6 \cdot 4^1$	0.131
7	$\binom{7}{7} \cdot 6^7 \cdot 4^0$	0.028

outcomes. We may depict this binomial probability distribution as in fig. A.5. When n is large, the calculation of binomial probabilities with the use of (A.9) requires a prohibitive amount of work. Tables are available for various values of θ and for large values of n. In point pattern analysis, however, the binomial distribution is not used as often as other distributions which are more easily evaluated. Nonetheless, the binomial does help in understanding one of these, the Poisson distribution.

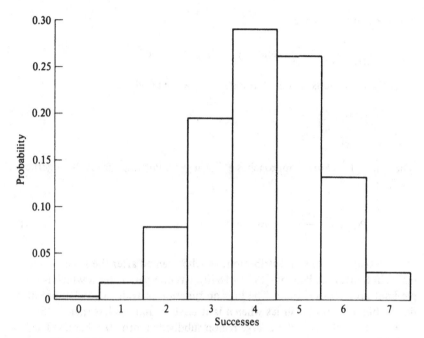

Fig. A.5 Binomial distribution ($n = 7$, $\theta = 0.6$)

A.6.2 The Poisson distribution

In many instances of map pattern analysis the value of n is very large and θ is very small. The binomial distribution becomes very difficult to deal with in those circumstances and thus the Poisson distribution is used as a very good approximation of binomial probabilities.

n is usually large because samples of size 10 or 20 are less acceptable than those of greater size. Small θ results from experiments where there are many possible outcomes other than those labeled success (the rest being considered failures). For example, consider the large number of possible locations for points on a map.

We shall show how the binomial is related to the Poisson distribution. Let n become very large ($n \to \infty$) and θ become very small ($\theta \to 0$) while $n\theta$ itself remains a moderate magnitude. Let $n\theta = \lambda$. Also assume that x is very small compared with n. First we have the binomial

$$P(x; n, \theta) = \binom{n}{x} \theta^x (1 - \theta)^{n-x}$$

which is approximately equal to

$$\frac{n^x}{x!} \theta^x (1 - \theta)^n$$

as $n \to \infty$ and $\theta \to 0$ since

$$\frac{n!}{x!(n-x)!} \simeq \frac{n^x}{x!} \quad \text{and} \quad (1-\theta)^{n-x} \simeq (1-\theta)^n$$

under the conditions specified. Substituting λ for $n\theta$ yields

$$\frac{\lambda^x \left(1 - \frac{\lambda}{n}\right)^n}{x!}$$

The value $[1 - (\lambda/n)]^n$ approaches $e^{-\lambda}$ as $n \to \infty$. Putting our results together we find

$$P(x; \lambda) = \frac{\lambda^x e^{-\lambda}}{x!} \quad \text{for } x = 0, 1, 2, \ldots \tag{A.10}$$

Note that the Poisson distribution, which is named after the French mathematician Simeon D. Poisson (1781–1840), has one parameter λ which is the pre-limit product of n and θ. The Poisson function provides a good approximation to binomial probabilities when n is at least 20 and θ is less than 0.05.

As an example, consider a large region subdivided into 20 subareas. The probability that a particular point falls into a particular subarea is 1/20 or $\theta = 0.05$. If there are 48 points to be placed in the region then $n\theta = \lambda = 2.4$. Suppose we wish to know the probability of any subarea containing three points, then

$$P(3; 2.4) = \frac{(2.4)^3 e^{-2.4}}{3!} = 0.209$$

The corresponding binomial probability is 0.215.

The complete distribution for the above example is given in table A.3 which also contains for comparison the case when $n = 48, \theta = 0.025$ and $n\theta = 1.2$. In chapter 2 the Poisson distribution is shown to have many important applica-

Table A.3. *Examples of Poisson distributions,* $\lambda = 2.4$ *and* $\lambda = 1.2$

Number of points contained in area (x)	Probability distribution	
	$P(x; 2.4)$	$P(x; 1.2)$
0	0.091	0.301
1	0.218	0.361
2	0.261	0.217
3	0.209	0.087
4	0.125	0.026
5	0.060	0.006
$\geqslant 6$	0.036	0.002
	1.000	1.000

tions. Some of these are results of what are called Poisson process models which can be derived independently of the binomial distribution.

A.7.   Moments of probability distributions

Very often it is necessary to compute the mean, variance and other characteristics of a probability distribution. Computation of these values some-times makes it unnecessary to go through the process of calculating each prob-ability ($P(x)$) of a random variable.

The mean of a random variable is obtained by summing the product of each of two elements — the value of the random variable and its associated probability. This is

$$\mu = \Sigma x \cdot P(x) \tag{A.11}$$

The mean (μ) of the random variable shown in table A.4 is 1.25. This value is

Table A.4. *Calculation of* μ_1

x	$p(x)$	$x \cdot p(x)$
0	0.38	0.00
1	0.25	0.25
2	0.18	0.36
3	0.12	0.36
4	0.07	0.28
	1.00	1.25

also called the *mathematical expectation* of a discrete random variable X and it is written

$$E(X) = \mu$$

The variance is a measure of the spread or dispersion of a distribution in relation to the mean. The variance is called the *second moment about the mean*. There is a distinction between moments about the origin and moments about the mean.

The first moment is the mean itself and can be written

$$\mu_1 = \mu = E(X) = \Sigma x \cdot P(x) \tag{A.12}$$

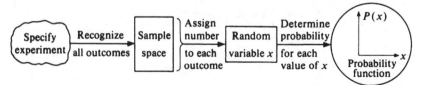

Fig. A.6 Steps in the construction of a probability model. Source: Harnett (1975, p. 88)

Subsequent moments about the mean can be found from the general formula

$$\mu_r = \Sigma(x - \mu)^r P(x) \quad \text{for } r = 1, 2, \ldots, n \tag{A.13}$$

Based on the above formula we calculate the variance, μ_2, from the distribution shown in table A.4.

$$\mu_2 = \Sigma(x - 1.25)^2 P(x)$$

$$= (0 - 1.25)^2(0.38) + (1 - 1.25)^2(0.25) + (2 - 1.25)^2(0.18)$$

$$+ (3 - 1.25)^2(0.12) + (4 - 1.25)^2(0.07)$$

$$= 1.6075$$

The lower the value of μ_2 the less spread the distribution. Relative to the mean, high variance values indicate dispersed probability distributions.

   The third moment about the mean (μ_3) is sometimes used to describe the symmetry or lack of symmetry (skewness) of a distribution. For the above distribution $\mu_3 = +1.4288$ which implies that there is an elongation in the positive direction (skewed to the right). A value of 0 implies perfect symmetry.

A.7.1.   Moments of the binomial distribution

   To determine the mean of the binomial distribution from (A.11), we evaluate the sum

$$\mu_1 = \sum_{x=0}^{n} x \cdot P(x; n, \theta)$$

$$= \sum_{x=0}^{n} x \binom{n}{x} \theta^x (1 - \theta)^{n-x}$$

$$= \sum_{x=1}^{n} \frac{n!}{(x-1)!(n-x)!} \theta^x (1 - \theta)^{n-x}$$

where the term for which $x = 0$ is omitted because it is equal to zero. Note that x divided by $x!$ is $1/(x-1)!$. We can then factor out the n in $n!$ and θ and thus obtain

$$\mu_1 = n\theta \sum_{x=1}^{n} \binom{n-1}{x-1} \theta^{x-1} (1 - \theta)^{n-x}$$

The value

$$\sum_{x=1}^{n} \binom{n-1}{x-1} \theta^{x-1} (1 - \theta)^{n-x}$$

is itself a binomial distribution which like all other probability distributions sums to one, thus

$$\mu_1 = n\theta \tag{A.14}$$

Without going through the derivation, the value for the variance is

$$n\theta(1-\theta) \tag{A.15}$$

From table A.2 we get

$$\mu_1 = n\theta = (7)(.6) = 4.2$$

and $\quad \mu_2 = n\theta(1-\theta) = 1.68$

The third moment about the mean is

$$\mu_3 = n\theta(1-\theta)(1-2\theta) = 1.075 \tag{A.16}$$

A.7.2. Moments of the Poisson distribution

By applying the same limiting process $(n \to \infty, \theta \to 0)$ to the mean and the variance of the binomial distribution we can obtain the mean and variance of the Poisson distribution. Since $\mu_1 = n\theta = \lambda$ and $\mu_2 = n\theta(1-\theta) = \lambda(1-\theta) \to \lambda$ when $\theta \to 0$ we have

$$\mu_1 = \lambda \tag{A.17}$$

and $\quad \mu_2 = \lambda \tag{A.18}$

The fact that the first two moments are the same is important in the analysis of map patterns. Very often a quick comparison of the mean and variance of an observed distribution can tell much about the nature of the pattern.

Appendix B

Table of critical values of D in the Kolmogorov–Smirnov one-sample test*

Sample size (N)	Level of significance for D = maximum ($\lvert$observed − expected$\rvert$)				
	0.20	0.15	0.10	0.05	0.01
1	0.900	0.925	0.950	0.975	0.995
2	0.684	0.726	0.776	0.842	0.929
3	0.565	0.597	0.642	0.708	0.828
4	0.494	0.525	0.564	0.624	0.733
5	0.446	0.474	0.510	0.565	0.669
6	0.410	0.436	0.470	0.521	0.618
7	0.381	0.405	0.438	0.486	0.577
8	0.358	0.381	0.411	0.457	0.543
9	0.339	0.360	0.388	0.432	0.514
10	0.332	0.342	0.368	0.410	0.490
11	0.307	0.326	0.352	0.391	0.468
12	0.295	0.313	0.338	0.375	0.450
13	0.284	0.302	0.325	0.361	0.433
14	0.274	0.292	0.314	0.349	0.418
15	0.266	0.283	0.304	0.338	0.404
16	0.258	0.274	0.295	0.328	0.392
17	0.250	0.266	0.286	0.318	0.381
18	0.244	0.259	0.278	0.309	0.371
19	0.237	0.252	0.272	0.301	0.363
20	0.231	0.246	0.264	0.294	0.356
25	0.21	0.22	0.24	0.27	0.32
30	0.19	0.20	0.22	0.24	0.29
35	0.18	0.19	0.21	0.23	0.27
Over 35	$\dfrac{1.07}{\sqrt{N}}$	$\dfrac{1.14}{\sqrt{N}}$	$\dfrac{1.22}{\sqrt{N}}$	$\dfrac{1.36}{\sqrt{N}}$	$\dfrac{1.63}{\sqrt{N}}$

* Adapted from Massey, F. J. Jr (1951) The Kolmogorov–Smirnov test for goodness of fit, *Journal of the American Statistical Association*, **46**, 70, with the kind permission of the author and publisher.

Glossary of general notation

$\binom{n}{x}$ the binomial coefficient (see appendix section A.3).

D_{obs} the calculated Kolmogorov–Smirnov value for a given sample.

$D_{0.20}$ the critical Kolmogorov–Smirnov value for the 0.20 significance level for a given sample size.

df   degrees of freedom.

$E(\ )$ expected value. Values that would be expected to result, on the average, from the random variable described by the probability function (see appendix section A.7). Measures which summarize probability distributions.

$F(x)$ cumulative probability distribution of a random variable x.

$f(x)$ probability density function of a continuous random variable x.

$g(\ )$ gamma distribution. A continuous probability distribution described in section 3.2.1.

Γ symbol used to refer to gamma function. For any positive integer k,

$$\Gamma(k) = (k-1)!, \text{ for positive real values of } k, \Gamma(k) = \int_0^\infty x^{k-1} e^{-x}\, dx.$$

lower-case Greek letters such as $\alpha, \delta, \gamma, \lambda$ parameters of various probability functions.

$\hat{\alpha}, \hat{\delta}, \hat{\gamma}, \hat{\lambda}$, etc.   estimates of parameters of probability functions.

m   the mean of a sample frequency distribution.

μ the expected or mean value of a random variable (see appendix section A.7).

μ_j the jth moment about the mean (see appendix section A.7).

O_i, f_i the observed frequency of items in the ith category.

E_i the expected frequency of items in the ith category.

p a probability.

p_{ij}, p_x the probabilities associated with the ij^{th}, x categories, respectively.

$P(\ )$ the probability associated with the category denoted within the parentheses. More often it is used to represent the various probabilities of an entire probability distribution.

$\sigma^2, v(x)$ the variance of a random variable.

σ the standard deviation of a random variable.

v the variance of a sample frequency distribution.

w, x, y, etc.   random variables (see appendix section A.6). x is also used for a location in the plane.

w_i, x_i, y_i the ith values of the random variables.

χ^2 the chi-squared distribution.

X^2 a test statistic which is approximately distributed as χ^2.

Z the standard variate of the normal curve.

Bibliography

Abramowitz, M. and Stegun, I. A. (eds.) (1965), *Handbook of mathematical functions*, New York: Dover Publications.

Aldskogius, H. (1969), Modelling the evolution of settlement patterns: Two studies of vacation house settlement, *Geografiska Regionstudier*, (6), Uppsala.

Anderson, D. L. (1972), A simple spatial diffusion model, *in* Adams, W. P. and Helleiner, F. M. (eds.), *International Geography 1972*, Papers of the International Geographical Congress, University of Toronto Press, 947–8.

Anscombe, F. J. (1949), The statistical analysis of insect counts based on the negative binomial distribution, *Biometrics*, 5, 165–73.

Anscombe, F. J. (1950), Sampling theory of the negative-binomial and logarithmic series distributions, *Biometrika*, 37, 358–82.

Archibald, E. E. A. (1950), Plant populations. II. The estimation of the number of individuals per unit area of species in heterogeneous plant populations, *Annals of Botany, London*, N.S., 12, 221–35.

Armstrong, R. A. (1974), Dynamics of expanding inhibitory fields, *Science*, 183, 444–5.

Artle, R. K. (1965), *The Structure of the Stockholm economy*, American Edition. Ithaca, New York: Cornell University Press.

Bartholomew, D. J. (1967), *Stochastic models for social processes*, New York: John Wiley.

Bartlett, M. S. (1964), The spectral analysis of two-dimensional point processes, *Biometrika*, 51, 299–311.

Batty, M. (1974), Spatial entropy, *Geographical Analysis*, 4(1), 1–31.

Beard, C. N. (1959), Quantitative study of columnar jointing, *Bulletin of the Geological Society of America*, 70, 379–82.

Berry, B. J. L. and Marble, D. F. (eds.) (1968), *Spatial analysis: A reader in statistical geography*, Englewood Cliffs, New Jersey: Prentice-Hall.

Berry, B. J. L. and Schwind, P. J. (1969), Information and entropy in migration flows, *Geographical Analysis*, 1, 5–14.

Blacklith, R. E. (1958), Nearest-neighbor distance measurements for the estimation of animal populations, *Ecology*, 39, 150–7.

Bliss, C. I. (1953), Fitting the negative binomial distribution to biological data, *Biometrics*, 9, 176–96.

Bliss, C. I. and Owen, A. R. G. (1958), Negative binomial distribution with a common *k*, *Biometrika*, 45, 37–58.

Bogue, D. J. (1949), *The structure of the metropolitan community: a study of dominance and subdominance*, Ann Arbor: Horace M. Rackham School of Graduate Studies, University of Michigan.

Boots, B. N. (1970), An approach to the study of patterns of cellular nets, *Rutgers University, Department of Geography, Discussion Papers*, 1.

Boots, B. N. (1972), The study of cellular nets, unpublished Ph.D. dissertation, Rutgers University.

Boots, B. N. (1973), Some models of the random subdivision of space, *Geografiska Annaler*, **55B**, 34–48.

Boots, B. N. (1974), Delaunay triangles: an alternative approach to point pattern analysis, *Proceedings of the Association of American Geographers*, **6**, 26–9.

Boots, B. N. (1975a), Patterns of urban settlements revisited, *The Professional Geographer*, **27**, 426–31.

Boots, B. N. (1975b), Some observations on the structure of socio-economic cellular networks, *The Canadian Geographer*, **19**, 107–20.

Boots, B. N., Getis, A. and Hagevik, G. (1972), Evaluation of an air pollution intensity model in the northeast New Jersey–New York city area, *Geographical Analysis*, **4**, 373–91.

Boots, B. N. and High, C. J. (1974), Some observations on the nature of clustering in point patterns, *Northeast Regional Science Review*, **4**, 144–52.

Boots, B. N. and Merk, G. G. (1974), People on beaches: some indications of the utility of a search for microlevel location theory, *Proceedings of the Middle States Division, Association of American Geographers*, **7**, 41–5.

Bosch, A. J. (1963), The Polya distribution, *Statistica Neerlandica*, **17**, 201–13.

Boswell, M. T. and Patil, G. P. (1970), Chance mechanisms generating the negative binomial distribution, *in* Patil, G. P. (ed.), *Random counts in scientific work*, vol. 1, 3–22.

Brown, L. A. (1963), The diffusion of innovation: a Markov chain-type approach, *Department of Geography, Northwestern University, Discussion Papers*, 3.

Brown, L. A. (1965), Models for spatial diffusion research, *Office of Naval Research, Geography Branch, Task 389–140, Technical Report*, 3.

Brown, L. A. (1968), *Diffusion processes and location theory*, Regional Science Research Institute, Bibliography Series, No. 4.

Brown, L. A. (1970), On the use of Markov chains in movement research, *Economic Geography*, **46(2)** (supplement), 393–403.

Bunge, W. (1966), *Theoretical geography*, Revised edition. Lund Studies in Geography, Series C, General and Mathematical Geography, 1.

Chapman, G. P. (1970), The application of information theory to the analysis of population distributions in space, *Economic Geography*, **46(2)**, 317–31.

Chapman, G. P. (1973), The spatial organization of the population of the United States and England and Wales, *Economic Geography*, **49(4)**, 325–43.

Christofides, N. and Eilon, S. (1969), Expected distances in distribution problems, *Operational Research Quarterly*, **20**, 437–43.

Clark, P. J. (1956), Grouping in spatial distributions, *Science*, **123**, 373–4.

Clark, P. J. and Evans, F. C. (1954), Distances to nearest neighbor as a measure of spatial relationships in populations, *Ecology*, **35**, 445–53.

Clark, P. J. and Evans, F. C. (1955), On some aspects of spatial pattern in biological populations, *Science*, **121**, 397–8.

Clark, W. A. V. (1969), Applications of spacing models in intra-city studies, *Geographical Analysis*, **1**, 391–9.

Clarke, D. L. (1972), A provisional model of Iron Age society and its settlement system, *in* Clarke, D. L. (ed.), *Models in archaeology*, London: Methuen.

Cliff, A. D. and Ord, J. K. (1973), *Spatial autocorrelation*, London: Pion.

Cohen, A. C. Jr. (1960), Estimating the parameter in a conditional Poisson distribution, *Biometrics*, **16**, 203–11.

Cohen, J. E. (1971), *Casual groups of monkeys and men: stochastic models of elemental social systems*, Cambridge, Mass.: Harvard University Press.

Coleman, S. R. (1964), *Introduction to mathematical sociology*, New York: Free Press.

Cooper, L. (1974), A random locational equilibrium problem, *Journal of Regional Science*, **14**, 47–54.

Cowie, S. R. (1968), The cumulative frequency nearest neighbour method for the identification of spatial patterns, *University of Bristol, Department of Geography, Seminar Series A*; No. 10.

Cox, D. R. and Miller, H. D. (1965), *The theory of stochastic processes*, New York: John Wiley.

Cox, K. R. and Agnew, J. A. (1974), Optimal and non-optimal territorial partitions: a possible approach toward conflict, *Papers of the Peace Science Society (International)*, **23**, 123–38.

Crain, I. K. (1972), Monte Carlo simulation of the random Voronoi polygons: preliminary results, *Search*, **3(6)**, 220–1.

Cross, C. A. and Fisher, D. L. (1968), The computer simulation of lunar craters, *Monthly Notices of the Royal Astronomical Society*, **139**, 261–72.

Curry, L. (1964), The random spatial economy: an exploration in settlement theory, *Annals of the Association of American Geographers*, **54**, 138–46.

Curry, L. (1972), A spatial analysis of gravity flows, *Regional Studies*, **6**, 131–47.

Dacey, M. F. (1960a), A note on the derivation of nearest neighbor distances, *Journal of Regional Science*, **2**, 81–7.

Dacey, M. F. (1960b), The spacing of river towns, *Annals of the Association of American Geographers*, **50**, 59–61.

Dacey, M. F. (1962), Analysis of central place and point pattern by a nearest neighbor method, *Lund Studies in Geography*, **24**, 55–75.

Dacey, M. F. (1963a), Order neighbor statistics for a class of random patterns in multidimensional space, *Annals of the Association of American Geographers*, **53**, 505–15.

Dacey, M. F. (1963b), Certain properties of edges on a polygon in a two dimensional aggregate of polygons having randomly distributed nuclei, unpublished paper, Wharton School of Finance and Commerce, University of Pennsylvania (Mimeographed).

Dacey, M. F. (1964a), Two-dimensional random point patterns: A review and an interpretation, *Papers, Regional Science Association*, **13**, 41–55.

Dacey, M. F. (1964b), Modified Poisson probability law for point pattern more regular than random, *Annals of the Association of American Geographers*, **54**, 559–65.

Dacey, M. F. (1965), Order distance in an unhomogeneous random point pattern, *The Canadian Geographer*, **9**, 144–53.

Dacey, M. F. (1966a), A compound probability law for a pattern more dispersed than random and with areal inhomogeneity, *Economic Geography*, 42, 172–9.

Dacey, M. F. (1966b), A county seat model for the areal pattern of an urban system, *Geographical Review*, 56, 527–42.

Dacey, M. F. (1966c), A probability model for central place location, *Annals of the Association of American Geographers*, 56, 550–68.

Dacey, M. F. (1967), Description of line patterns, *Northwestern Studies in Geography*, 13, 277–87.

Dacey, M. F. (1968), An empirical study of the areal distribution of houses in Puerto Rico, *Transactions, Institute of British Geographers*, 45, 15–30.

Dacey, M. F. (1969a), Proportion of reflexive n^{th} order neighbors in spatial distribution, *Geographical Analysis*, 1(4), 385–8.

Dacey, M. F. (1969b), A hypergeometric family of discrete probability distributions: Properties and applications to location models, *Geographical Analysis*, 1, 283–317.

Dacey, M. F. (1969c), Some properties of a cluster point process, *Canadian Geographer*, 13, 128–40.

Dacey, M. F. (1969d), Some spacing measures of areal point distributions having the circular normal form, *Geographical Analysis*, 1, 15–30.

Dacey, M. F. (1969e), Similarities in the areal distributions of houses in Japan and Puerto Rico, *Area*, 3, 35–7.

Dacey, M. F. (1973), A central focus cluster process for urban dispersion, *Journal of Regional Science*, 13(1), 77–90.

Dacey, M. F. and Tung, T.–H. (1962), The identification of randomness in point patterns, *Journal of Regional Science*, 4, 83–96.

DeVos, S. (1973), The use of nearest neighbor methods, *Tijdschrift voor Economische en Sociale Geografie*, 64(5), 307–19.

Dirichlet, G. L. (1850), Uber die Reduction der positeven quadratischen. Formen mit drei unbestimmten ganzen Zahlen, *Journal für die reine und angewandte Mathematik*, 40, 209–27.

Douglas, J. B. (1955), Fitting the Neyman Type A (two parameters) contagious distribution, *Biometrics*, 11, 149–73.

Eilon, S., Watson-Gandy, C. D. T. and Christofides, N. (1970), *Distribution management: mathematical modeling and practical analysis*, London: Griffin.

Evans, U. R. (1945), The laws of expanding circles and spheres in relation to the lateral growth of surface films and the grain-size of metals, *Transactions of the Faraday Society*, 41, 365–74.

Feller, W. (1943), On a general class of contagious distributions, *Annals of Mathematical Statistics*, 14, 389–400.

Feller, W. (1968), *An introduction to probability theory and its application*, vol. 1, 3rd edition, New York: John Wiley.

Fielder, G. (1966), Tests for randomness in the distribution of lunar craters, *Monthly Notices of the Royal Astronomical Society*, 132, 413–22.

Fielder, G. and Marcus, A. (1967), Further tests of randomness in lunar craters, *Monthly Notices of the Royal Astronomical Society*, 136, 1–10.

Fisher, R. A. (1941), The negative binomial distribution, *Annals of Eugenics*, 11, 182–7.

Fisher, R. A. (1953), Note on the efficient fitting of the negative binomial distribution, *Biometrics*, 9, 197–200.

Freund, J. E. (1971), *Mathematical statistics*, 2nd edition, Englewood Cliffs, New Jersey: Prentice-Hall.

Garbrecht, D. (1971), Pedestrian paths through a uniform environment, *Town Planning Review*, **42**, 71–84.

Garrison, C. B. and Paulson, A. S. (1973), An entropy measure of the geographic concentration of economic activity, *Economic Geography*, **49(4)**, 319–24.

Garwood, F. and Holroyd, E. M. (1966), The distance of a 'random chord' of a circle from the centre, *Mathematical Gazette*, **50**, 283–6.

Getis, A. (1963), The determination of the location of retail activities with the use of a map transformation, *Economic Geography*, **39**, 14–22.

Getis, A. (1964), Temporal land use pattern analyses with the use of the nearest neighbor and quadrat methods, *Annals of the Association of American Geographers*, **54**, 391–8.

Getis, A. (1967), Occupancy theory and map pattern analysis, *Seminar Paper Series, Department of Geography, University of Bristol, Series A*, No. 1.

Getis, A. (1969), Some thoughts on a negative binomial model and geographic data, *Quantative Methods in Geography: A Symposium*, American Geographical Society, Mimeographed and Offset Publication No. 6, 31–43.

Getis, A. (1974), Representation of spatial point pattern processes by Polya models, *in* Yeates, M. H. (ed.), *Proceedings of the 1972 meeting of the IGU commission on quantitative geography*, Montreal: McGill–Queen's University Press, 76–100.

Getis, A. and Jackson, P. H. (1971), The expected proportion of a region polluted by *k* sources, *Geographical Analysis*, **3**, 256–61.

Getis, A. and Merk, G. (1973), Spacing of human groups, *Proceedings, Association of American Geographers*, **5**, 80–3.

Gilbert, E. N. (1961), Random plane networks, *SIAM Journal of Applied Mathematics*, **9**, 533–43.

Gilbert, E. N. (1962), Random subdivisions of space into crystals, *Annals of Mathematical Statistics*, **33**, 958–72.

Gilbert, E. N. (1965), Random minimal trees, *SIAM Journal of Applied Mathematics*, **13(2)**, 376–87.

Glass, L. (1973), Stochastic generation of regular distributions, *Science*, **180**, 1061–3.

Glass, L. and Tobler, W. R. (1971), Uniform distribution of objects in a homogeneous field: cities on a plain, *Nature*, **233**, 67–8.

Graustein, W. C. (1931), On the average number of sides of polygons for a net, *Annals of Mathematics*, **32(2)**, 149–53.

Greig-Smith, P. (1952), The use of random and contiguous quadrats in the study of the structure of plant communities, *Annals of Botany, London*, N.S., **16**, 293–316.

Greig-Smith, P. (1964), *Quantitative plant ecology*, 2nd edition, London: Butterworths.

Gurland, J. (1957), Some interrelations among compound and generalized distributions, *Biometrika*, **44**, 265–8.

Gurland, J. (1958), A generalized class of contagious distributions, *Biometrics*, **14**, 229–49.

Hägerstrand, T. (1952), The propagation of innovation waves, *Lund Studies in Geography, Series B*, **4**, 3–19.

188   Models of spatial processes

Hägerstrand, T. (1953), *Innovationsförloppet ur Korologisk synpunkt*, Lund University, Lund.
Hägerstrand, T. (1957), Migration and area, *in* Hannenberg, O., Hägerstrand, T. and Odeving, B., *Migration in Sweden*, Lund Studies in Geography, Series B, 13.
Hägerstrand, T. (1967), *Innovation diffusion as a spatial process*, University of Chicago Press.
Haggett, P. (1965), *Locational analysis in human geography*, London: Edward Arnold.
Haggett, P. (1967), Network models in geography, *in* Chorley, R. J. and Haggett, P. (eds.), *Models in geography*, London: Methuen, 609–68.
Haggett, P. and Chorley, R. J. (1969), *Network analysis in geography*, London: Edward Arnold.
Haight, F. A. (1959), The generalized Poisson distribution, *Annals of the Institute of Statistical Mathematics*, Tokyo, 11, 101–5.
Haight, F. A. (1967), *Handbook of the Poisson distribution*, New York: John Wiley.
Hamilton, W. D. (1971a), Geometry for the selfish herd, *Journal of Theoretical Biology*, 31, 295–311.
Hamilton, W. D. (1971b), Selection of selfish and altruistic behavior in some extreme models, *in* Eisenberg, J. F. and Dillon, W. S. (eds.), *Man and beast: comparative social behavior*, Washington, D.C.: Smithsonian Institution Press.
Hammersley, J. M. (1972), Stochastic models for the distribution of particles in space, *Advances in Applied Probability* (Supplement), 47–68.
Hammond, N. D. C. (1972), Locational models and the site of Lubaantua: a classic Maya centre, *in* Clarke, D. L. (ed.), *Models in archaeology*, London: Methuen.
Harnett, D. L. (1975), *Introduction to statistical methods*, 2nd edition, Reading, Mass.: Addison-Wesley.
Harvey, D. W. (1966), Geographic processes and the analysis of point patterns: Testing a diffusion model by quadrat sampling, *Transactions and Papers, Institute of British Geographers*, 40, 81–95.
Harvey, D. W. (1967), Models of the evolution of spatial patterns in human geography, *in* Chorley, R. J. and Haggett, P. (eds.), *Models in geography*, London: Methuen.
Harvey, D. W. (1968), Some methodological problems in the use of the Neyman Type A and the negative binomial probability distributions for the analysis of spatial point patterns, *Transactions and Papers, Institute of British Geographers*, 44, 85–95.
Haynes, K. E. and Enders, W. T. (1975), Distance, direction, and entropy in the evolution of a settlement pattern, *Economic Geography*, 51, 357–65.
Hinz, P. and Gurland, J. (1967), Simplified techniques for estimating parameters of some generalized Poisson distributions, *Biometrika*, 54, 555–66.
Hodder, I. R. (1972a), The interpretation of spatial patterns in archaeology: two examples, *Area*, 4(4), 223.
Hodder, I. R. (1972b), Locational models and the study of Romano-British settlement, in Clarke, D. L. (ed.), *Models in archaeology*, London: Methuen.
Holgate, P. (1965), Tests of randomness based on distance methods, *Biometrika*, 52, 345–53.

Holgate, P. (1972), The use of distance methods for the analysis of spatial distribution of points, *in* Lewis, P. A. W. (ed.), *Stochastic point processes: statistical analysis, theory, and applications.* New York: Wiley-Interscience, 122–35.

Horowitz, M. (1965), Probability of random paths across elementary geometrical shapes, *Journal of Applied Probability*, 2, 169–77.

Hsu, S. and Mason, J. D. (1974), The nearest-neighbor statistics for testing randomness of point distributions in a bounded two-dimensional space, *in* Yeates, M. H. (ed.), *Proceedings of the 1972 meeting of the IGU commission on quantitative geography*, Montreal: McGill–Queen's University Press, 32–54.

Hudson, J. (1969), A location theory for rural settlement, *Annals of the Association of American Geographers*, 59(2), 365–81.

Hudson, J. (1973), Density and pattern in suburban fringes, *Annals of the Association of American Geographers*, 63(1), 28–39.

Irwin, J. O. (1959), On the estimation of the mean of a Poisson distribution from a sample with the zero class missing, *Biometrics*, 15, 324–6.

Jackson, J. L. (1974), Stochastic generation of regular distributions: some analytical results, *Science*, 183, 445–6.

James, G. and James, R. C. (eds.), (1949), *Mathematics dictionary*, New York: Van Nostrand.

James, J. (1951), A preliminary study of the size determinant in small group interaction, *American Sociological Review*, 16, 474–7.

James, J. (1953), The distribution of free-forming small group size, *American Sociological Review*, 18, 569–70.

Johnson, N. L. and Kotz, S. L. (1969), *Discrete distributions*, Boston: Houghton Mifflin.

Johnson, W. A. and Mehl, R. F. (1939), Reaction kinetics in processes of nucleation and growth, *Transactions of the American Institute of Mining, Metallurgical and Petroleum Engineers*, 135, 410–58.

Jones, A. (1971), An order neighbor approach to random disturbances on regular point lattices, *Geographical Analysis*, 4, 361–78.

Katti, S. K. and Gurland, J. (1962), Efficiency of certain methods of estimation for the negative binomial and the Neyman Type A distributions, *Biometrika*, 49, 215–26.

Kemeny, J. G. and Snell, J. L. (1960), *Finite Markov chains*, Princeton: Van Nostrand.

Kemp, C. D. and Kemp, A. W. (1956), The analysis of point quadrat data, *Australian Journal of Botany*, 4, 167–74.

Kendall, D. G. (1949), Stochastic processes and population growth, *Journal of the Royal Statistical Society, Series B*, 11, 230–64.

Kendall, M. G. and Moran, P. A. P. (1963), *Geometrical probability*, Griffin's Statistical Monographs and Courses, No. 5, London: C. Griffin.

Kendall, M. G. and Buckland, W. R. (eds.), (1972), *A dictionary of statistical terms*, Edinburgh: Oliver and Boyd.

Kiang, T. (1966), Random fragmentation in two and three dimensions, *Zeitschift für Astrophysik*, 64, 433–9.

King, L. J. (1961), A multivariate analysis of the spacing of urban settlements in the United States, *Annals of the Association of American Geographers*, 51, 222–33.

King, L. J. (1962), A quantitative expression of the pattern of urban settlements in selected areas of the United States, *Tijdschrift voor Economische en Sociale Geografie*, 53, 1–7.

King, L. J. (1969), *Statistical analysis in geography*, Englewood Cliffs, New Jersey: Prentice-Hall.

Lee, Y. (1972), A stochastic model of the geometric patterns of urban settlements and urban spheres of influence: A clumping model, *Geographical Analysis*, 4(1), 51–64.

Lee, Y. (1974), An analysis of spatial mobility of urban activities in downtown Denver , *The Annals of Regional Science*, 8(1), 95–108.

Lenz, R. D. (1977), Information theory and change in map patterns, Ph.D. dissertation, Rutgers University.

Lewis, F. T. (1946), The shape of cells as a mathematical problem, *American Scientist*, 34, 359–69.

Ling, R. F. (1972), On the theory and construction of *k*-clusters, *The Computer Journal*, 15(4), 326–32.

Ling, R. F. (1973*a*), A probability theory of cluster analysis, *Journal of the American Statistical Association*, 68, 159–64.

Ling, R. F. (1973*b*), The expected number of components in random linear graphs, *Annals of Probability*, 1, 876–81.

MacDougall, E. B. (1972), Optimal generalization of mosaic maps, *Geographical Analysis*, 4(4), 416–23.

Mack, C. (1954), The expected number of clumps when convex laminae are placed at random and with random orientation on a plane area, *Proceedings of the Cambridge Philosophical Society*, 50, 581–5.

Mack, C. (1956), On clumps formed when convex laminae are placed at random in two or three dimensions, *Proceedings of the Cambridge Philosophical Society*, 52, 246–50.

Marcus, A. (1966), A stochastic model of the formation and survival of lunar craters. IV. On the non-randomness of crater centres, *Icarus*, 5, 190–200.

Marcus, A. (1967), A multivariate immigration with multiple death process and applications to lunar craters, *Biometrika*, 54, 251–61.

Martin, D. C. and Katti, S. K. (1965), Fitting of certain contagious distributions to some available data by the maximum likelihood method, *Biometrics*, 21, 34–48.

Martin, J., Gomàr, N. and Rogers, A. (1969), *Some computer programs for quadrat analysis*, Working Paper, Center for Planning and Development Research, University of California, Berkeley.

Massam, B. H. (1974), A note on expected distances, *Geographical Analysis*, 6, 377–81.

Massam, B. H. (1975), *Location and space in social administration*, London: Edward Arnold.

Matérn, B. (1960), Spatial variation, *Meddelanden Fran Statens Skogsforsknings-institut*, 49(5), 1–144.

Matérn, B. (1971), Doubly stochastic Poisson processes in the plane, *in* Patil, G. P., Pielou, E. C. and Waters, W. E. (eds.), *Statistical ecology*, vol. 1, 193–213, Pennsylvania State University Press.

Matschinski, M. (1954), Considération statistique sur les polygones et les polyèdres, *Publications of the Institute of Statistics, University of Paris*, 3, 179–201.

Matui, I. (1932), Statistical study of the distribution of scattered villages in two regions of the Tonami Plain, Toyama Prefecture, *Japanese Journal of Geology and Geography*, **9**, 251–6.

McConnell, H. (1966), Quadrat methods in map analysis, *Department of Geography, University of Iowa, Discussion Paper*, **3**.

McFadden, J. A. (1965), The entropy of a point process, *Journal of the Society for Industrial and Applied Mathematics*, **13(4)**, 988–94.

Mead, R. (1971), Models for interplant competition in irregularly distributed populations, *in* Patil, G. P., Pielou, E. C. and Waters, W. E. (eds.), *Statistical ecology*, vol. 2, Pennsylvania State University Press.

Medvedkov, Y. V. (1967), Concept of entropy in settlement pattern analysis, *Papers, Regional Science Association*, **18**, 165–8.

Medvedkov, Y. V. (1970), Entropy: an assessment of potentialities in geography, *Economic Geography*, **46(2)**, 306–16.

Meijering, J. L. (1953), Interface area, edge length, and number of vertices in crystal aggregates with random nucleation, *Philips Research Reports*, **8**, 270–90.

Miles, R. E. (1964*a*), Random polygons determined by random lines in a plane: I, *Proceedings of the National Academy of Sciences*, **52**, 901–7.

Miles, R. E. (1964*b*), Random polygons determined by random lines in a plane: II, *Proceedings of the National Academy of Sciences*, **52**, 1157–60.

Miles, R. E. (1970), On the homogeneous planar Poisson point process, *Mathematical Biosciences*, **6**, 85–127.

Moore, P. J. (1954), Spacing in plant populations, *Ecology*, **35**, 222–7.

Morisita, M. (1954), Estimation of population density by spacing method, *Memoirs of the Faculty of Science, Kyushu University*, E, I, 187–97.

Morrill, R. L. (1963), The development of spatial distributions of towns in Sweden: an historical-predictive approach, *Annals of the Association of American Geographers*, **53**, 1–14.

Morrill, R. L. (1965), Migration and the spread and growth of urban settlement, *Lund Studies in Geography, Series B*, **26**.

Naus, J. I. (1965), Clustering of random points in two dimensions, *Biometrika*, **52**, 263–7.

Naus, J. I. and Rabinowitz, L. (1975), The expectation and variance of the number of components in random linear graphs, *The Annals of Probability*, **3**, 159–61.

Neft, D. S. (1966), *Statistical analysis for areal distributions*, Monograph series, 2, Philadelphia: Regional Science Research Institute.

Neyman, J. (1939), On a new class of 'contagious' distributions, applicable in entomology and bacteriology, *Annals of Mathematical Statistics*, **10**, 35–57.

Neyman, J. and Scott, E. L. (1958), Statistical approach to problems of cosmology, *Journal of the Royal Statistical Society*, B, **20**, 1–43.

Norcliffe, G. B. (1967), Areal grouping with elementary nearest neighbour linkage analysis, *University of Bristol, Department of Geography, Seminar Paper Series A;* No. 12.

Norcliffe, G. B. (1968), Matrix analysis of changing patterns of industrial location, *University of Bristol, Department of Geography, Seminar Paper Series A*; No. 16.

Nordbeck, S. (1965), The law of allometric growth, *Michigan Inter-University Community of Mathematical Geographers, Discussion Paper*, **7**.

Olsson, G. (1965), *Distance and human interaction: a review and bibliography*, Bibliography Series, No. 2; Regional Science Research Institute, Philadelphia.

Olsson, G. and Gale, S. (1968), Spatial theory and human behavior, *Papers of the Regional Science Association*, 21, 229–42.

Ord, J. K. (1972), *Families of frequency distribution*, London: Griffin.

Ord, J. K. and Patil, G. P. (1972), Lectures on ecological problems and statistical distributions, The Advanced Institute on Statistical Ecology in the United States, The Pennsylvania State University. (Mimeographed).

Patil, G. P. (ed.) (1965), *Classical and contagious discrete distributions*, New York: Pergamon Press.

Patil, G. P. (ed.) (1970), *Random counts in scientific work*, 3 vols., Pennsylvania State University Press.

Patil, G. P., Pielou, E. C. and Waters, W. E. (eds.) (1971), *Statistical ecology*, 3 vols., Pennsylvania State University Press.

Pedersen, P. O. (1967), On the geometry of administrative areas Copenhagen, M. S. Report.

Persson, O. (1971), The robustness of estimating density by distance measurements, *in* Patil, G. P., Pielou, E. C. and Waters, W. E. (eds.), *Statistical ecology*, vol. 2, Pennsylvania State University Press, 175–90.

Pielou, E. C. (1957), The effect of quadrat size on the estimation of the parameters of Neyman's and Thomas' distributions, *Journal of Ecology*, 45, 31–47.

Pielou, E. C. (1959), The use of point-to-point populations, *Journal of Ecology*, 47, 607–13.

Pielou, E. C. (1961), Segregation and symmetry in two-species populations as studied by nearest-neighbor relations, *Journal of Ecology*, 49, 255–69.

Pielou, E. C. (1962), The use of plant-to-neighbor distances for the detection of competition, *Journal of Ecology*, 50, 357–67.

Pielou, E. C. (1965), The concept of randomness in the patterns of a mosaic, *Biometrics*, 21(4), 908–20.

Pielou, E. C. (1967), A test for random mingling of the phases of a mosaic, *Biometrics*, 23(4), 657–70.

Pielou, E. C. (1969), *An introduction to mathematical ecology*, New York: John Wiley & Sons.

Pinder, D. A. and Witherick, M. E. (1972), The principles, practice and pitfalls of nearest-neighbour analysis, *Geography*, 57, 277–88.

Porter, P. W. (1960), Earnest and the Orephagians – a fable for the instruction of young geographers, *Annals of the Association of American Geographers*, 50, 297–9.

Quenouille, M. H. (1949), A relation between the logarithmic, Poisson, and negative binomial series, *Biometrics*, 5, 162–4.

Raisbeck, G. (1963), *Information theory: an introduction for scientists and engineers*, Boston: Massachusetts Institute of Technology Press.

Rhynsburger, D. (1973), Analytic delineation of Thiessen polygons, *Geographical Analysis*, 5(2), 133–44.

Roach, S. A. (1968), *The theory of random clumping*, London: Methuen's Monographs on Applied Probability and Statistics.

Roberts, F. D. K. (1967), A Monte Carlo solution of a two-dimensional unstructured cluster problem, *Biometrika*, 54, 625–8.

Roberts, F. D. K. (1968), Random minimal trees, *Biometrika*, 55, 255–8.

Roder, W. (1974), Application of a procedure for statistical assessment of points on a line, *The Professional Geographer*, 26, 283–90.

Roder, W. (1975), A procedure for assessing point patterns without reference to area or density, *The Professional Geographer*, 27, 432–40.

Rogers, A. (1965), A stochastic analysis of the spatial clustering of retail establishments, *Journal of the American Statistical Association*, 60, 1094–103.

Rogers, A. (1968), *Matrix analysis of interregional population growth and distribution*, Berkeley: University of California.

Rogers, A. (1969a), Quadrat analysis of urban dispersions: 1. Theoretical techniques, *Environment and Planning*, 1, 47–80.

Rogers, A. (1969b), Quadrat analysis of urban dispersion: 2. Case studies of urban retail systems, *Environment and Planning*, 1, 155–71.

Rogers, A. (1974), *Statistical analysis of spatial dispersion*, London: Pion.

Rogers, A. and Gomar, N. (1969), Statistical inference in quadrat analysis, *Geographical Analysis*, 1(4), 370–84.

Rogers, A. and Martin, J. (1971), Quadrat analysis of urban dispersion: 3. Bivariate models, *Environment and Planning*, 3, 433–50.

Rogers, A. and Raquillet, R. (1972), Quadrat analysis of urban dispersion: 4. Spatial sampling, *Environment and Planning*, 4(3), 331–45.

Rushton, G. (1971), Map transformations of point patterns: central place patterns in areas of variable population density, Discussion paper No. 17, Department of Geography, The University of Iowa.

Santalo, L. A. and Yanez, I. (1972), Averages for polygons formed by random lines in Euclidean and hyperbolic planes, *Journal of Applied Probability*, 9, 140–57.

Schilling, W. (1947), A frequency distribution represented as the sum of two Poisson distributions, *Journal of the American Statistical Association*, 42, 407–24.

Scott, A. J. (1971), *Combinational programming, spatial analysis and planning*, London: Methuen.

Semple, R. K. (1973), Recent trends in the spatial concentration of corporate headquarters, *Economic Geography*, 49(4), 309–18.

Semple, R. K. and Golledge, R. G. (1970), An analysis of entropy changes in a settlement pattern over time, *Economic Geography*, 46, 157–60.

Shaw, R. P. (1975), *Migration theory and fact*, Regional Science Research Institute, Bibliography Series, No. 5.

Shenton, L. R. (1949), On the efficiency of the method of moments and Neyman's Type A distribution, *Biometrika*, 36, 450–4.

Siegel, S. (1956), *Nonparametric statistics for the behavioral sciences*, New York: McGraw-Hill.

Skellam, J. G. (1948), A probability distribution derived from the binomial distribution by regarding the probability of success as variable between the sets of trials, *Journal of the Royal Statistical Society, B*, 10B, 257–61.

Skellam, J. G. (1953), Studies in statistical ecology – 1. Spatial pattern, *Biometrika*, 39, 346–62.

Skellam, J. G. (1958), On the derivation and applicability of Neyman's Type A distribution, *Biometrika*, 45, 32–6.

Smalley, I. J. (1966), Contraction crack networks in basalt flows, *Geological Magazine*, 103(2), 110–14.

Smalley, I. J. (1970), The two-dimensional distribution of some geomorphological features: a simple random placement model, Paper presented at the autumn meeting of the British Geomorphological Research Group, University of Cambridge, 31 October 1970.

Smith, C. S. (1952), Grain shape and other metallurgical applications of topology. Metal Interfaces, *American Society for Metals*, Cleveland, 65–108.

Smith, C. S. (1954), The shape of things, *Scientific American*, **190**, 58–64.

Sorensen, A. D. (1974), A method for measuring the spatial association between point patterns, *Professional Geographer*, **26(2)**, 172–6.

Susling, Jr., W. G. (1971a), Analyzing the pattern of discrete spatial distributions: selected topics in near neighbor analysis, *Department of Geography, Indiana University, Discussion Paper Series*, 2.

Susling, Jr., W. G. (1971b), Analyzing the pattern of discrete spatial distributions: selected topics in quadrat analysis, *Department of Geography, Indiana University, Discussion Paper Series*, 3.

Tarver, J. D. and Gurley, W. R. (1965), A stochastic analysis of geographic mobility and population projections of the census divisions of the United States, *Demography*, **2**, 134–9.

Taylor, P. J. (1971), Distances within shapes: an introduction to a family of finite frequency distributions, *Geografiska Annaler*, **53** B(1), 40–53.

Theil, H. (1967), *Economics and information theory*, Amsterdam: North Holland Publishing Company.

Thomas, M. (1949), A generalization of Poisson's binomial limit for use in ecology, *Biometrika*, **36**, 18–25.

Thompson, H. R. (1956), Distribution of distance to n-th neighbor in a population of randomly distributed individuals, *Ecology*, **37**, 391–4.

Thompson, H. R. (1958), The statistical study of plant distribution patterns using a grid of quadrats, *Australian Journal of Botany*, **6**, 322–43.

Tinkler, K. J. (1972), Bounded planar networks: a theory of radial structures, *Geographical Analysis*, **4(1)**, 5–33.

Tinline, R. R. (1968), Towards a Markov diffusion model, *Department of Geography, University of Bristol, Seminar Paper Series A*, No. 13.

Tobler, W. R. (1963), Geographic area and map projections, *Geographical Review*, **53**, 59–78.

Tobler, W. R. (1970), *Selected computer programs*, Ann Arbor: Department of Geography, University of Michigan.

Vitek, J. D. (1973), Patterned ground: a quantitative analysis of pattern, *Proceedings of the Association of American Geographers*, **5**, 272–5.

Voronoi, G. (1908), Nouvelles applications des paramètres continus à la théorie des formes quadratiques, Deuxième Memoire, Recherches sur les parallelloèdres primitifs, *Journal für die reine und angewandte Mathematik*, **134**, 198–287.

Walsh, J. E. (1963), Bounded probability properties of Kolmogorov–Smirnov and similar statistics for discrete data, *Annals of the Institute of Statistical Mathematics*, **15**, 153–8.

Williamson, E. and Bretherton, M. H. (1963), *Tables of the negative binomial probability distribution*, New York: John Wiley & Sons, Inc.

Williamson, E. and Bretherton, M. H. (1964), Tables of the logarithmic series distribution, *Annals of Mathematical Statistics*, **35**, 284–97.

Woldenberg, M. J. (1970), The hexagon as a spatial average, *Harvard Papers in Theoretical Geography*, **42**.

Index

Abramowitz, M. 17
additive processes
    superimposition 38–45
    heterogeneity 45–8
agglomeration models 11
Agnew, J. A. 136
Aldskogius, H. 58, 67
analytic procedures 14–15
Anderson, D. L. 78
Anscombe, F. J. 52, 56, 66, 68
area patterns
    cell model 126–37
    clumping models 152–63
    Johnson–Mehl model 145–51
    processes for generating 121–3
    properties 123–6
Armstrong, R. A. 149
Artle, R. F. 55
assignment models 13, 132
association models 12

Bartholomew, D. J. 71
Bartlett, M. S. 63
Batty, M. 144
Beard, C. N. 136, 138
Berry, B. J. L. 33, 144
Bertrand, J. 98
binomial coefficients 169
binomial distribution 173, 178–9
Blacklith, R. E. 33
Bliss, C. I. 52, 66
Bogue, D. J. 132
Boots, B. N. 55, 63, 117, 129, 132, 137, 142, 149, 150, 151, 160, 161
Bosch, A. J. 76
Boswell, M. T. 56, 67
branching models 12
Bretherton, M. H. 55, 64
Brown, L. A. 16, 74, 75, 107
Bunge, W. 99
bus service center hinterlands 132–5

cell growth rate 145–51
cell models
    area pattern assumptions 126–37
    description 13
    growth interpretation 132–7, 145–51
    information theory and 142–4
    line pattern assumptions 107–16
    moments for properties 126–32
central places 9, 13, 133–4
Chapman, G. P. 143, 144
chi-square one sample test 22–3, 63
Chorley, R. J. 16, 104, 107, 116
Christofides, N. 98
circuit models 13, 104–7
Clark, P. J. 26, 32, 33, 115
Clark, W. A. V. 33
Clarke, D. L. 132
Cliff, A. D. 35, 67
clumping models
    the Getis–Jackson model 156–61
    modified model 162–3
    the Roach approach 152–6
cluster identification 63
Cohen, Jr., A. C. 35
Cohen, J. E. 36, 37
Coleman, S. R. 37
combinatorial theory 166–9
    combinations 168–9
    permutations 167–8
competition models 12
contagion effect 5, 6, 59, 71–8
contiguous patterns
    definition of 122
    processes generating 126–44
Cooper, L. 98
coverage models 13
Cowie, S. R. 31
Cox, D. R. 74
Cox, K. R. 136
Crain, I. K. 129, 130, 131

Cross, C. A. 163
Curry, L. 16, 144

Dacey, M. F. 16, 32, 33, 43, 44, 48, 55,
  63, 66, 67, 76, 78, 81, 112–16,
  118–20, 127, 128, 129
Dacey's tests (line patterns)
  nearest neighbor approach 112–14,
    118–19
  random walk procedure 115–16,
    119–20
  reciprocals approach 114–15
degrees of freedom 23
Delaunay triangles 137–42
DeVos, S. 33
diffusion models 11–12
Dirichlet, G. L. 128
Dirichlet regions 128
distance measures
  location-to-point 27
  point-to-point 26–7
  reciprocal or reflexive 32–3
  segregation and symetry 34–5
  significance tests 27–30
  to second, third, etc. nearest
    neighbors 30–2
distribution, definition of 16
disturbed lattice models 78–81
double Poisson process model 45–7
doubly stochastic Poisson process 63
Douglas, J. B. 57, 64

Eilon, S. 98
Enders, W. T. 85
Evans, F. C. 26, 32, 33
Evans, U. R. 127, 128, 145, 148
exodic tree 101

family development effect 5, 6
Feller, W. 16, 33, 64, 78
Fielder, G. 162
Fisher, D. L. 163
Fisher, R. A. 56
Freund, J. E. 23n
friction of distance effect 5

Gale, S. 74
gamma distribution 50–1, 131–2
gamma function 50
Garbrecht, D. 94
Garrison, C. B. 144
Garwood, F. 98
general double Poisson process model
  47–8, 70
Getis, A. 3, 4, 6, 33, 37, 55, 63, 64, 67,
  78, 81, 156–61
Getis–Jackson model 156–63

Gilbert, E. N. 99–102, 104–7, 127, 128,
  131, 147
Gilbert model 104–7
Glass, L. 149
Golledge, R. G. 85
Gomar, N. 25, 63, 64
Graustein, W. C. 129
Green, F. H. W. 132
Greig–Smith, P. 24, 25
growth models
  cell model 126–37
  description 13–14
  Johnson–Mehl model 145–51
Gurland, J. 47, 55. 56, 64
Gurley, W. R. 74

Hägerstrand, T. 61, 62, 67, 75
Hagevik, G. 161
Haggett, P. 16, 102, 104, 107, 116, 129
Haight, F. A. 33, 47
Hamilton, W. D. 63, 126
Hammond, N. D. C. 132
Harnett, D. L. 177
Harvey, D. W. 55, 58, 61, 62, 67, 68, 74
Haynes, K. E. 85
High, C. J. 55, 63
Hinz, P. 47
Hodder, I. R. 33, 132
Holgate, P. 33
Holroyd, E. M. 98
Horowitz, M. 93–8
Horowitz, model 93–8
Hsu, S. 33
Hudson, J. 16

index of dispersion 25–6
information gain 144
information theory
  and the cell model 142–4
  quadrat approach 81–5
Irwin, J. O. 36

Jackson, J. L. 149
Jackson, P. H. 63, 156–61
James, J. 37
Johnson, N. L. 16, 33, 37, 52, 55, 56,
  64, 65, 68, 78
Johnson, W. A. 145–51
Johnson–Mehl model 145–51
  Boots' modifications 149–51
  coefficients of variation 149
  correlation coefficients 149
  description and assumptions 145–7
  moments 147–9
Jones, A. 81

Katti, S. K. 64

Kemeny, J. G. 74
Kemp, A. W. 25
Kemp, C. D. 25
Kendall, D. G. 68
Kendall, M. G. 99, 136
Kiang, T. 126, 131, 132
King, L. J. 16, 33, 81
Kolmogorov–Smirnov one-sample test
    description 97–8
    table of critical values 180
Kotz, S. L. 16, 33, 37, 52, 55, 56, 64,
    65, 68, 78

Lee, Y. 55, 63, 157, 158, 159
Lenz, R. D. 143
line patterns
    processes for generating 86–92
    properties 92–3
Ling, R. F. 63, 106
linkage processes 12–13
location-to-point distances 27
logarithmic series distribution 64–6
lost cause effect 5, 6

MacDougall, E. B. 117
Mack, C. 159
Marble, D. F. 33
Marcus, A. 162
Markov–chain models 71–4
Martin, D. C. 64
Mason, J. D. 33
Massam, B. H. 98
Massey, Jr., F. J. 180
Matérn, B. 16, 63
mathematical expectation 177
Matschinski, M. 129
Matui, I. 33
maximum likelihood technique 52
McConnell, H. 55, 68
Medvedkov, Y. V. 85, 144
Mehl, R. F. 145–51
Meijering, J. L. 126, 127, 128, 132, 148
Merk, G. 37, 117
migration processes 5–6
Miles, R. E. 33, 108–12, 114–16, 120,
    139, 140, 141
Miles' model 108–12
Miller, H. D. 74
minimal tree 99
Mixed Poisson process models
    additive processes 38–48
    multiplicative processes 48–70
model assumptions 9–11
Moore, P. J. 33
Moran, P. A. P. 99, 136
Morisita, M. 33
Morrill, R. L. 75

multinomial coefficient 170
multiplicative processes
    compound models 48–56
    generalized models 56–68

Naus, J. I. 106
nearest neighbor measures
    general 26–35
    line patterns 112–14, 118–19
Neft, D. S. 63
negative binomial process model
    as a compound model 51–6
    as a generalized model 64–7
Neyman, J. 57, 63, 64
Neyman Type A process model 57–61
Norcliffe, G. B. 33, 74
Nordbeck, S. 157
nucleation rate 145, 150

occupancy theory 171
Olsson, G. 16, 74
Ord, J. K. 16, 35, 56, 64, 67
overlap pattern 122
Owen, A. R. G. 66

partitioning models 14
path models 12, 93–9
Patil, G. P. 17, 56, 67
pattern change 12
pattern, definition of 8–9
Paulson, A. S. 144
Pedersen, P. O. 129
Persson, O. 33
pied piper effect 5, 6
Pielou, E. C. 16, 17, 25, 27, 33, 34, 35,
    62, 63, 64, 66, 116–18, 126
Pinder, D. A. 33
point patterns
    measurement 20–35
    mixed Poisson process models 38–70
    Poisson process model 18–20
    truncated Poisson process model 35–7
point-to-point distances 26–7
Poisson, S. D. 176
Poisson distribution 18–20, 175–7, 179
Poisson plus Bernoulli process model
    42–5
Poisson plus Poisson process model
    40–2
Poisson process model
    assumptions of 18–20
    cluster process assumptions 58–9
    pattern measurement 20–35
    truncated model 35–7
pollution area model 160–2
Polya–Aeppli process model 67–70
Polya–Eggenberger process model 75–8

Porter, P. W. 33
probability distributions
    definitions of 171–3
    moments of 177–9

quadrat
    sampling 20–5
    size 23–5, 63
    tests 22–3, 25–6
Quenouille, M. H. 65

Rabinowitz, L. 106
random birth effect 6
random death effect 6
random, definition of 15–16
random variable 171
Raquillet, R. 64
rat trap effect 5, 6
reciprocal or reflexive measures
    general 32–3
    line patterns 114–15
Rhynsburger, D. 141
Roach approach 152–6
Roach, S. A. 63, 105, 152–6, 157, 159
Roberts, F. D. K. 99, 105
Roder, W. 31
Rogers, A. 16, 25, 55, 56, 58, 63, 64, 65, 68, 74
Rushton, G. 81

Santalo, L. A. 111
Schilling, W. 47, 70
Schwind, P. J. 144
Scott, A. J. 99
Scott, E. L. 63
segregation models 12
segregation or symmetry measures 34–5
Semple, R. K. 85, 144
set notation 165–6
settlement patterns 53–5, 157–9
Shaw, R. P. 75
Shenton, L. R. 64
Siegel, S. 98
simulation models 75
Skellam, J. G. 33, 56, 64
Smalley, I. J. 126, 136–8
Smith, C. S. 129
S–mosaic 126
Snell, J. L. 74

Sorenson, A. D. 33
space-exhaustive pattern 122–3
spatial autocorrelation 54–5
spatial processes
    definition of 1–3
    examples 11–14
    framework for viewing 4–9
    models of 9–11
spread effect 6
Stegun, I. A. 17
survival of the fittest effect 5, 6
Susling, Jr., W. G. 33, 55, 68

Tarver, J. D. 74
Taylor, P. J. 99
Theil, H. 144
Thiessen polygons 126–8, 135–42
Thomas, M. 62
Thomas process model 61–4
Thompson, H. R. 25, 30, 33
Tinkler, K. J. 102
Tinline, R. R. 74
Tobler, W. R. 81
town meeting effect 5, 6
transition probabilities 71–4
tree models 99–104
truly contagious models
    Markov–chain models 71–5
    simulation models 75
    urn models 75–8
truncated Poisson process model 35–7, 59–61
t test 25
Tung, T. –H. 81
two-phase mosaics 116–18

urn models 75–8

variance-mean ratio 25, 63
Venn diagrams 165–6
Voronoi, G. 128
Voronoi polygons 128

Walsh, J. E. 98
Waters, W. E. 17
Williamson, E. 55, 64
Witherick, M. E. 33
Woldenberg, M. J. 129

Yanez, I. 111